INDIANA'S WEATHER AND CLIMATE

Indiana Natural Science
Gillian Harris, editor

QUARRY BOOKS
AN IMPRINT OF
Indiana University Press
Bloomington and Indianapolis

INDIANA'S
WEATHER
AND
CLIMATE

JOHN E. OLIVER

This book is a publication of

Quarry Books
an imprint of

Indiana University Press
601 North Morton Street
Bloomington, IN 47404-3797 USA

http://iupress.indiana.edu

Telephone orders   800-842-6796
Fax orders   812-855-7931
Orders by e-mail   iuporder@indiana.edu

The paper used in this publication meets the minimum requirements of American National Standard for Information Sciences--Permanence of Paper for Printed Library Materials, ANSI Z39.48-1984.

Manufactured in China

Library of Congress Cataloging-in-Publication Data

Oliver, John E.
  Indiana's weather and climate / John E. Oliver.
    p. cm. — (Indiana natural science)
  Includes bibliographical references and index.
  ISBN 978-0-253-22056-1 (pbk. : alk. paper)
1. Weather. 2. Climatology. 3. Indiana—Climate. I. Title.
  QC981.O425 2009
  551.69772—dc22

                                     2008034834

1 2 3 4 5 14 13 12 11 10 09

CONTENTS

PREFACE

The remarkable technological advances over the last twenty-five years are reflected in our understanding of the atmosphere. Each day satellite views of the earth, color enhanced to show weather systems, are routinely seen on television screens; radar images provide instant images of current weather conditions and warnings of severe weather. Views of the earth's climate have also improved enormously using technology and we are now in a position to view the earth as the climate changes from season to season and year to year. As illustrated by El Niño, events that occur thousands of miles away impact the conditions that exist in the United States. The careful reconstruction of past conditions and the monitoring of current conditions have enabled the identification of increasing global temperatures, and global warming has become an event of political and economic concern. Using the speed of new computer systems, it has become possible to look at future climates using models which are becoming more and more sophisticated. In all, human ingenuity using the available tools has made the study of weather and climate a major scientific area of study.

While, to some, a quick glance at a television weather map, or a brief newspaper account of global warming, may be enough to satisfy curiosity, there remains in many of us a need to look a little closer at what the atmosphere is doing. This book provides material to meet that goal. To be sure, global warming and computer models are discussed, but so are woolly worms, degree days, and cooperative weather observers. It is possible for an account dealing with climate to contain endless lists of numbers, from daily records to monthly totals. While I offer a few illustrative tables for the reader, I have avoided a heavily quantitative presentation of the issues. Similarly, as far as possible, I have avoided using citations and references in the text and instead use the appendix to list them.

The complexity of the atmosphere requires the expertise of specialists. As the list of contributors shows, I have drawn freely upon many. I have also relied heavily upon the Indiana State Climate Office and its staff. Without their support a book on Indiana's weather and climate would prove impossible. I also appreciate the support of the Department of Geography at Indiana State University and its chair, Dr. Susan Berta.

Finally, and as ever, my wife, Loretta, provided the enthusiasm and encouragement necessary to complete this work.

ACKNOWLEDGMENTS

There are many experts who deal with the weather and climate of Indiana. This text draws upon the following scholars who aided significantly in the preparation of this volume. Each individual contribution is identified within the text.

Ronald L. Baker, Department of English, Indiana State University
Greg Bierly, Department of Geography, Indiana State University
Ashley Brooks, Indiana State Climate Office, Purdue University
Umarporn "Jam" Charusombat, Indiana State Climate Office, Purdue University
Craig A. Clark, Department of Geography and Meteorology, Valparaiso University
Cameron Craig, Department of Geology and Geography, Eastern Illinois University
Souleymane Fall, Indiana State Climate Office, Purdue University
J. A. Howe, Atmospheric Science Program, Department of Geography, Indiana University
Roger Kenyon, National Weather Service, Indianapolis
John Kwiatkowski, National Weather Service, Indianapolis
Kenneth E. Kunkel, Midwestern Regional Climate Center, University of Illinois
Chuck Lofton, Meteorologist, WTHR-TV, Indianapolis
Zack Payne, Indiana State Climate Office, Purdue University
S. C. Pryor, Atmospheric Science Program, Department of Geography, Indiana University
Ken Scheeringa, Indiana State Climate Office, Purdue University
A. M. Spaulding, Atmospheric Science Program, Department of Geography, Indiana University
James Speer, Department of Ecology and Environmental Sciences, Indiana State University
Steve Stadler, Department of Geography, Oklahoma State University and Oklahoma State Geographer

Many climate maps included in this work are derived from the *Climatic Atlas of Indiana* now being prepared by the Indiana State Climate Office at Purdue University. They are used with permission of its director, Dr. Dev Niyogi, whose aid in preparing this work was invaluable. His research associates are identified in the acknowledgement listing above.

I am greatly indebted to Cameron Craig of Eastern Illinois University for constructing many of the text figures. His work is also acknowledged in appropriate captions.

Professor S. C. Pryor of the Atmospheric Science Program at Indiana University kindly guided this manuscript through the copyediting and proofreading process after the untimely death of Professor Oliver.

INDIANA'S WEATHER AND CLIMATE

INTRODUCTION

The people of Indiana have long been concerned about the weather and climate they experience. Consider, for example, the June 11, 1835, edition of the *Wabash Courier* where the following account appeared.

> So far, the season has been one of the most extraordinary within our recollection. During the last twenty days we have had drenching rain—torrents of it—affecting the Wabash and smaller streams . . . On many farms the sprouting corn is nearly washed away . . . in the progress of the National road, operations are nearly suspended along the whole line.

With but a few changes, this account, written some 175 years ago, might easily be found in a newspaper of recent years.

While the accounts of weather events of past years may be compared with those of today, the modern Hoosier has access to much more information about atmospheric conditions. Radios provide hourly reports and almost instantaneous information when storms occur; television news shows provide satellite images and radar information as part of the daily weather coverage and forecast. The well-known Weather Channel even provides twenty-four-hour coverage. Such scientific information has largely replaced the weather lore that once prevailed. Blackberry Winters, which suggested that when frost lies on the blackberry blossoms it ensured a rich harvest, or the observation of the width of the brown band on the back of the woolly worm as a guide to the severity of winter, are but reminders of the way in which observations of nature were used for prediction.

There are many reasons for Hoosiers' long-held interest in weather and climate; with little doubt, the agricultural basis of the state's economy has been and is of prime importance. That agricultural production is largely successful when weather conditions are close to normal, and potentially disastrous when extremes occur, is testimony to the intimate relationship between Hoosiers and their environment. But agriculture is not the only arena in which the atmospheric conditions are significant. Climate and weather influence everyday life and livelihood in many other ways ranging from energy costs for heating and cooling, to relative comfort of the outdoor conditions. Such impacts are carefully examined in this book.

It will be noted that the volume title mentions both the *weather* and the *climate* of Indiana. There is a tendency to use the terms interchangeably, but in reality they refer to very different views of the atmosphere. Meteorology, the scientific study of weather, deals with motions and phenomena of the atmosphere with a view to forecasting the weather. It deals with conditions in the short term. Climatology, the study of climate, concerns atmospheric conditions over a period of at least thirty years. It deals not only with the average weather at a location, but also with extremes, variations, and change. As we all know, global climate change is frequently debated these days. The following illustrates the different types of questions dealt with by those studying weather and climate.

Questions That Might Be Asked about Weather and Climate

Weather	Climate
Will it rain tomorrow?	Is this the wettest month since 1900?
Will it freeze tonight?	What is the average date of the first frost?
What is the chance of severe weather from this storm system?	How many tornadoes were there in 1962?
What is the outlook for warm weather this weekend?	Can you predict average temperature for the year 2020?
Will today's weather be like that of yesterday?	What was the climate like 10,000 years ago?

The type of information provided by meteorologists and climatologists illustrates just how different are the two approaches. The weather expert will provide such facts as the high and low temperatures for today, or whether severe weather will occur within the next few hours. The climate specialist will comment on whether a recent drought is as severe as that which occurred in earlier years, or which area of the state has the highest frequency of tornadoes over the past fifty years. Both the weather and climate of Indiana are considered in this volume.

Given the abundance of information available today, it might well be asked why a book on Indiana's weather and climate would be useful. The answer, of course, rests with the fact that although we are provided by modern media with information about what the atmospheric conditions are like and where they might occur, we are seldom really told why they will, or have, happened. The chapters of this book provide not only the "what?" and "where?" of Indiana's weather and climate, but also the "why?"

In choosing the best way to organize all of the weather and climate information available, it seems clear that one way is to use one of Indiana's outstanding attributes, the changing seasons. I am sure that if one speaks to a former Hoosier, now living in Florida or perhaps Arizona, one will always hear from them "I miss the seasons." Each of the seasons has its own character and, to each of us, special meanings. The cold of winter is tempered by the beauty of a new snowfall, the heat of summer by the shade of Hoosier woodlands and forests.

Thus it is that, before an overview of the basic driving forces that produce the seasons, the book examines the basic characteristics of spring, summer, fall, and winter in Indiana.

They are considered in a number of ways, not the least important of which are the benefits and hazards that each season brings. The relative impacts of a late spring frost or the lack of summer rain are to many urban dwellers an inconvenience; to many Hoosier farmers, however, they can produce economic hardship. To some farmers a heavy winter snowfall is a positive feature because it provides moisture to the ground; to urban dwellers, the result can be a commuting nightmare.

Of course, in using the seasons as a basic guide, it needs to be asked how seasons are defined. Commonly, and as emphasized by the media, astronomical dates—that is, timing relative to the solstices—are used. As discussed in chapter 1, however, this division does not lend itself to climate analysis because climatologists group their data into four seasons based upon months.

Once the seasonal features are considered, attention turns to looking more closely at ways in which weather and climate are applicable to Indiana. Of some importance is the way that information and data are gathered for use in both short- and long-term atmospheric studies. For the short term, the compilation of the daily weather map and daily forecast are described. Thereafter, the emphasis is on climate: following an analysis of the present climate, the past and future climates are discussed. We then examine the way in which Indiana may be affected in the context of global warming.

A single author cannot always do justice to any set of events, so in this book you will find significant portions written by other atmospheric scientists. In some instances the contribution is a vignette and is presented as an *Of Interest* textbox within a chapter. In other instances, a specialist provides detailed information. In all, the aim in calling upon various experts is to provide the most up-to-date, authoritative, and clearly expressed account of the weather and climate of Indiana as is possible.

Indiana has rich and varied natural and human environments, and its weather and climate form a significant part of this milieu. If this text can add further to the understanding and appreciation of natural events in Indiana, then it will have achieved its goal.

WHY THE SEASONS?

The changes that occur from season to season provide one of Indiana's most beautiful natural attributes. The first blooms of spring, the warmth of summer, the changing leaves of fall, and the snow and frost of winter are well-known features of each year. In this chapter a number of the basic causes of seasons and seasonal differences are presented. Additionally, it shows how the natural transition from season to season is a result of the earth's position and movement in relation to our sun. This chapter, then, provides the basic background material for the analysis of Indiana's weather and climate that is dealt with in the chapters that follow.

Sun and Earth

The earth's atmosphere may be likened to a huge heat engine which, like any engine, requires an energy source to make it work. Our sun, a star composed of the gases hydrogen and helium, is the earth's energy source. In its core a nuclear reaction, in which hydrogen atoms are converted to helium, provides energy to the sun's radiating surface, which has a temperature of some 10,500°F (6000°K). At this temperature the sun radiates energy mostly in the form of light. The earth, some 93 million miles from the sun, intercepts but a tiny fraction of the total energy output of the sun; fortunately, the amount is sufficient to support life on earth while, at the same time, providing the necessary energy to drive the atmosphere.

The solar energy intercepted by the earth is not evenly distributed over its surface. This uneven distribution results in well-known differences between the generally warm tropical areas and the cooler polar realms. The temperature variations that occur are the result of a number of factors, including:

1) The time of year. Over the period of a year the position of the earth in relation to the sun causes changes in both the length of day and night, and the strength of the sun's rays.

2) The state of the sky. Heavily clouded skies modify the flow of energy to the surface and reduce the amount of available energy.

Days, Nights, and the Sun in the Sky

Residents of Indiana are well aware of the marked changes in the length of day and night throughout the year. Locations in the northern part of the state get a little more than nine hours of daylight on a December day but more than fifteen hours on a day in June. The southern part of the state receives somewhat fewer than fifteen hours on summer days but almost ten hours on the shortest December day.

The short period during which the sun is in the sky during the winter results in fewer sunlight hours and cooler temperatures. In contrast, the long summer days allow the surface to absorb much more energy and warmer temperatures prevail. This difference is enhanced in that during the winter months the sun is much lower in the sky than it is in summer. Figure 1.1 explains why this is so. When the sun is high in the sky the energy contained in a given amount of sunlight is more concentrated (that is, more intense) than when a low-angle sun prevails. In fact, the intensity of the sun's rays on a summer day is almost twice as great as on a day in winter.

The highest angle attained by the sun at noon on any day depends upon one's latitude. In the central part of Indiana, at around latitude 40°N, the highest angle occurs on June 21 when the midday sun is 73½° above the horizon. On December 21, the shortest day, the sun angle is only 26½°.

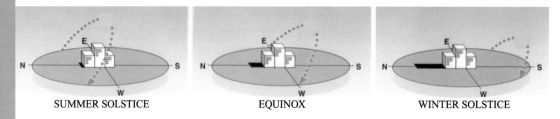

FIGURE 1.1. The height of the sun in the sky varies over the year. At the equinoxes, the sun rises in the east and sets in the west, and day and night are of equal length. The highest and lowest positions are at the summer and winter solstice, respectively. Figure courtesy of *NEBRASKAland Magazine*.

As solar energy flows through the atmosphere a number of things happen to modify the actual amount that eventually reaches the surface. Some of the short-wave energy (such as dangerous ultraviolet radiation in the UV-B and UV-C wavelengths) is absorbed by the ozone layer while much is absorbed and reflected by other atmospheric constituents. Often less than half of the solar beams arriving at the outer limits of the atmosphere reach the surface of the earth. Cloud cover is of particular importance in modifying the flow of solar energy to the earth's surface. Common everyday experience indicates how significant this can be. On a day when the sky is covered by a thick, continuous layer of clouds, a general gloom prevails. The exclusion of the sun's direct rays allows us only to perceive light that has been scattered by the overcast sky. A clear day, on the other hand, allows the sun's rays direct access to the surface. Temperature is the most easily sensed result of the changes in solar radiation receipts.

Temperatures

Temperature is one of the best-known and most widely used of all atmospheric variables. Indiana is in the middle latitude portion of the earth and its temperatures are neither excessively hot nor extremely cold for extended periods. It has a climate that is seasonal, and average yearly temperatures, as seen in figure 1.2, reflect this climate type.

It does need be noted that because temperature is merely a means of describing the heat contained by a given body, it is possible to express it on an arbitrary scale. The following note on temperature scales outlines this characteristic.

The fact that temperature is merely an arbitrary measure of heat means that it is measuring a form of energy. Heat energy is derived from solar energy which, as we know, is essentially in the form of light. The processes involved in obtaining heat energy from solar energy concern two main exchanges. First, solar energy striking a surface

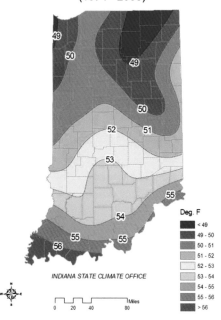

MEAN ANNUAL TEMPERATURE
(1974 - 2003)

INDIANA STATE CLIMATE OFFICE

Deg. F

■	< 49
■	49 - 50
■	50 - 51
■	51 - 52
□	52 - 53
■	53 - 54
■	54 - 55
■	55 - 56
■	> 56

FIGURE 1.2. The average annual temperature in Indiana, 1974–2003. Figure courtesy of the Indiana State Climate Office.

is absorbed by that surface, and the absorption of energy raises its temperature. The surface that absorbs the energy now becomes a radiating surface; the type of energy it radiates is a function of its temperature, and, in the ranges of temperature that occur at the earth's surface, it radiates infrared or heat energy. Thus, solar energy does not directly heat the lower atmosphere; rather, the absorbed energy radiated from the earth's surface is the prime source of heat. This important process leads to what is called the Greenhouse Effect in which the lower atmosphere is likened to a greenhouse being heated by solar energy. Figure 1.3 illustrates the Greenhouse Effect.

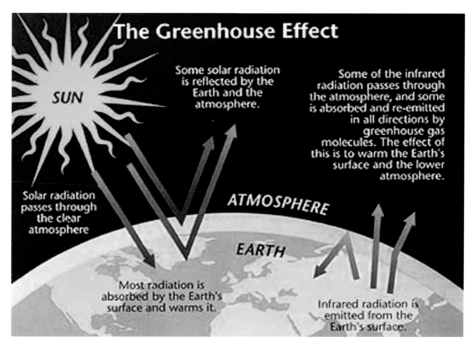

FIGURE 1.3. The Greenhouse Effect results from gases in the lower atmosphere causing part of earth's radiation to be delayed in its passage back to space. This results in a warming of earth's lower atmosphere. Figure courtesy of the National Oceanic and Atmospheric Administration (NOAA) and the National Aeronautics and Space Administration (NASA).

As expected, temperatures are highest during the day when large amounts of energy are flowing to earth and lowest at night. If a weather station has a self-recording thermometer, the temperature at each hour of the day can be recorded. An approximation of the daily mean temperature can be obtained as the average of the maximum and minimum temperatures that occur on a given day. The daily temperature is used to derive the average monthly temperatures for a given year by summing the values and dividing the number by the days in the month. Thereafter, by averaging the monthly temperature over a thirty-year period, the normal monthly temperature is derived. These monthly values may be used to derive a picture of the seasonal temperatures. As shown in later chapters, average temperatures show July as the warmest month.

Climatological Seasons

In using the seasons as a basic guide, it needs to be asked how seasons are defined. Commonly, and as emphasized by the media, astronomical dates are used. In this, the first day of summer is on or about June 21, and the first day of winter is on or about December 21. These are the dates when the apparent movement of the sun places it over the Tropic of Cancer (summer) and the Tropic of Capricorn (winter). Spring and fall are defined at the two times of the year when the sun is directly over the equator (on or about March 21, for spring, and September 21, for fall). To the climatologist, such a division leaves much to be desired. Winter weather occurs long before December 21, while June 21 is a date well into the warm summer season. Summarizing seasonal data using mid-month and variable dates also complicates analysis. As a result, climatologists group their data into four seasons based upon months.

Winter comprises December, January, and February

Spring comprises March, April, and May

Summer comprises June, July, and August

Fall comprises September, October, and November

This is the seasonal calendar used in this work.

Clouds

Clouds provide the most readily visible weather phenomena. Their beauty is not overlooked. Poems, songs, and paintings use clouds in highly expressive ways. While atmospheric scientists also see beauty in clouds, they look to the form and appearance of clouds as important keys to both understanding and predicting atmospheric conditions.

Before the year 1800, clouds had no formal names and there was little knowledge of cloud mechanics. It remained for a young Englishman named Luke Howard (1772–1864) to provide a new perspective on clouds. In 1803 he presented a classification of clouds into main and secondary types and gave them Latin names. He distinguished three principal cloud forms:

Stratus (from Latin stratum = layer) clouds—lying in a level sheet

Cumulus (from Latin cumulus = pile) clouds—having flat bases and rounded tops, and lumpy in appearance

Cirrus (from Latin = hair) cloud—having a fibrous or feathery appearance

The classification is so well designed that it remains the basic system in use today. However, as knowledge of clouds increased, a more comprehensive system was required. The international standard is now under the auspices of the World Meteorological Organization (WMO) which publishes *The International Cloud Atlas*.

Despite the infinite variety of clouds that occur, they group into one of ten basic types or genera. Clouds of similar shapes occur at different levels in the troposphere (that is, the earth's lowest atmospheric layer) and are grouped as high, middle, and low clouds. To some extent, the names of clouds relate to their height. Those with the term "cirrus" (or prefix "cirro-") are high clouds. Those with the prefix "alto" are middle clouds. Names for low clouds lack prefixes. An exception to this is the nimbostratus cloud classifies as both a middle and low cloud. The word "nimbus" (or prefix "nimbo-") applies to a cloud from which rain is falling. It derives from the Latin for "violent rain."

Figure 1.4 is a generalized diagram illustrating the classification of clouds according to basic types and height. Note that cumulonimbus, the thunder cloud, extends from low, through middle, to the high elevations of the height classification.

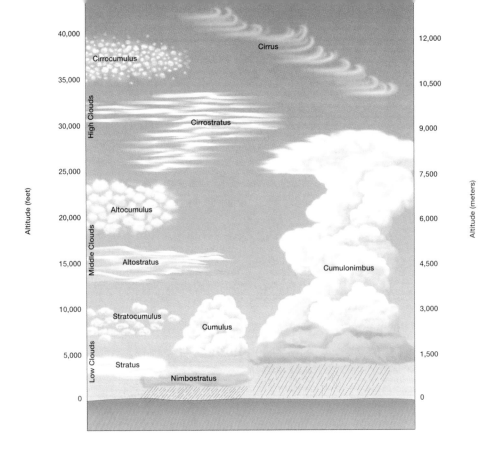

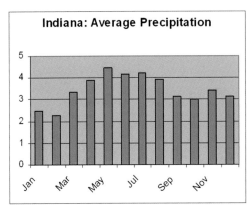

FIGURE 1.4. *(above)*
The ten principal types of clouds. Figure reprinted from McKnight (1996) by permission of Pearman Education, Inc.

FIGURE 1.5. *(left)*
The average monthly precipitation (inches) for all of Indiana. The spring-summer maximum is clearly seen. Figure courtesy of the Indiana State Climate Office.

Precipitation

The term "precipitation" is a useful one for the atmospheric scientist, for it is a general descriptor of all types of moisture coming to the earth from the atmosphere, from rainfall to snow, from hail to mist. The specialized precipitation types associated with varied cloud forms described above is considered in other chapters. At this point, only the general precipitation patterns in Indiana are pointed out.

Indiana is most fortunate because in most years there is sufficient precipitation for very productive agriculture and an ample water supply for cities and towns. As Figure 1.5 illustrates, for the state as a whole, every month has on average more than 2 inches of precipitation. May, June, and July are seen as receiving the most rainfall, closely followed by April and August.

Precipitation is not evenly distributed through the state. Figure 1.6 is a map of the mean annual precipitation for the state. The wettest part of the state is in the southern half, with more than 42 inches of precipitation annually. North of that, the yearly average is mostly between 36 and 42 inches with just one small eastern area receiving an average less than 36 inches of rain a year.

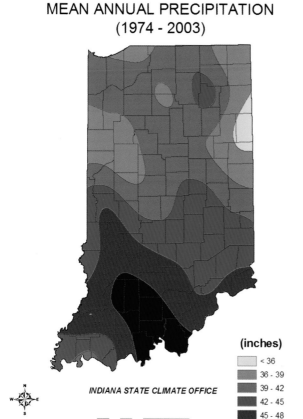

MEAN ANNUAL PRECIPITATION
(1974 - 2003)

INDIANA STATE CLIMATE OFFICE

(inches)
- < 36
- 36 - 39
- 39 - 42
- 42 - 45
- 45 - 48
- > 48

FIGURE 1.6.
The average annual precipitation in Indiana, 1974–2003. Figure courtesy of the Indiana State Climate Office.

Air in Motion

Moving air, the wind, results from differences in pressure that occur from place to place. Such pressure differences are caused by temperature variations and the dynamic effects of moving air itself. Like water that moves from high to low places, so air moves from high pressure to low pressure. And just as the rapidity of flowing water depends upon the slope, or gradient, of a hill, so wind speed depends upon the gradient between the high and low pressure centers. When the pressure gradient is very steep, as in the case of a tornado or a hurricane, then very strong winds occur.

In 1805 British Admiral Sir Francis Beaufort devised a practical method for estimating the force of wind at sea. He provided a scale ranging from 0 to 12 using common terms that sailors could understand. For example, Force 11 on his scale was "that which would produce a man-of-war to storm-stay sails." Such a description would not be of great value in Indiana. Instead we can use the Beaufort Scale that has been adapted for land observations. As seen in Table 1.1, if you can hear telephone wires whistling, then a Force 6 wind of 25–31 mph is blowing.

Table 1.1. Beaufort Wind Scale

Beaufort Number	Wind Speed (mph)	Description	Observed Effects (Land)
0	< 1	Calm	Calm; smoke rises vertically
1	1–3	Light air	Direction of wind shown by smoke drift, but not by wind vanes; slight leaf movement
2	4–7	Light breeze	Wind felt on face; leaves rustle; vanes moved by wind
3	8–12	Gentle breeze	Leaves and small twigs in constant motion; wind extends light flag
4	13–18	Moderate breeze	Raises dust, loose paper; small branches move
5	19–24	Fresh breeze	Small trees in leaf begin to sway; crested wavelets form on inland waters
6	25–31	Strong breeze	Large branches in motion; whistling heard in telephone wires; umbrellas used with difficulty
7	32–38	Near gale	Whole trees in motion; resistance felt when walking against wind
8	39–46	Gale	Breaks twigs off of trees; progress impeded while walking; autos drift
9	47–54	Strong gale	Slight structural damage occurs; branches down
10	55–63	Storm	Trees uprooted; considerable damage occurs
11	64–72	Violent storm	Widespread damage to trees and structures
12	> 73	Hurricane	Severe to catastrophic damage

The winds at the surface are only part of the great wind systems of the earth; for a more complete understanding the winds aloft need to be understood as well. Indiana is located in the middle latitudes, a region of the world dominated by planetary-scale westerly winds that blow from west to east. The weather and climate of this region are dominated by the upper air Westerlies, winds that occur at elevations between 5000 and 10,000 meters (about 16,000 to 33,000 feet). The Westerlies are the winds that act as a steering mechanism for the movement of great air masses and surface weather systems. Let us first look at the air masses.

Air Masses and Fronts

The great air masses are sorted by the areas in which they originate. During the winter season, Arctic regions receive no input of solar radiation because the sun remains below the horizon over vast areas. Without a heating source, the ground becomes extremely cold and chills the air above. A vast pool of air exists over frigid land and frozen sea, a pool that can be considered to be *continental* (abbreviated *c*) in origin. The great mass of air being chilled in Arctic (*A*), or polar (*P*), realms eventually earns the designation *continental polar* (*cP*) or *continental Arctic* (*cA*).

The pressure differences that occur within the air mass, together with prevailing circulation patterns, mean that it cannot remain stagnant forever. Eventually the cold air moves away from the source region. The direction and rate of movement is, as might be expected, controlled by patterns of the general circulation of the atmosphere, of which this cold air mass is an integral part. Outbursts of cold air are guided by the meandering upper air Westerlies, with a jet stream at its leading edge. As described in chapter 5, very rapid out-movement may give rise to an Alberta Clipper.

The cold air moving southward is not moving into a vacuum; it is displacing warmer air that probably originated over the sea (*m* for maritime) in tropical areas (*T* for tropics). The *mT* air is warmer and moister than the invading *cP* air, and along the line of demarcation, called a cold front, differences in the characteristics of the air masses appear. A cold front—shown in figure 1.7—is an area of conflict between two air masses, and the warmer air, as it is thrust upward, will give rise to heavy precipitation, often in the form of thundershowers. As the frigid continental air advances, it comes in contact with warmer and warmer ground, and eventually it is so modified that it more resembles the air into which it is moving than the source from which it came.

In summer, a main air mass influencing Indiana originates over the warm waters of the Gulf of Mexico or the Atlantic Ocean. This maritime (*m*) tropical (*T*) air brings the hot, sticky days of summer. If one of these *mT* air masses is advancing, and colder air is retreating, a warm front is formed. As seen in figure 1.8, this is a low-angle front and clouds form along it to cover large areas. The various types of air masses provide various amounts of moisture and eventually precipitation.

The analysis of fronts, together with an understanding of traveling high and low pressure systems, led to the identification of the middle latitude cyclone. A cyclone is a low pressure area with an organized circulation pattern of winds. In middle latitudes, cyclones develop at air mass boundaries and are characterized at the surface by the formation of fronts. A number of stages have been identified. The first stage initiates the formation of the central low pressure area and the two fronts. This is followed by an open stage in which the cold and warm fronts have formed with a distinctive warm sector between them. (An illustration of this is seen in the inset sections of figures 1.7 and 1.8.) Further evolution of the system leads to its dissolving state and the end of its life cycle. Each cyclone system is guided by the upper air winds across the United States, the Westerlies.

The Westerlies

The key to movement of air masses and the formation of fronts is the mid-latitude upper air Westerlies. As they encircle the globe, the upper air winds undergo changes. Initially, the winds blow almost directly from west to east with but a few slight variations. This zonal flow changes over time to form a wave-like pattern, known as meridional flow. In meridional flow, the waves form ridges of high pressure and troughs of low pressure. At the earth's surface, the resulting pattern is reflected in the type of air mass that will be present in a location.

When there is zonal flow, arctic air and tropical air are separated, and there is cold air to the north and warm weather south of the polar front. Since the dominate wind direction is west to east with zonal flow, air from the Pacific Ocean flows over the Rocky Mountains and crosses the continent. In the process of crossing the mountain ranges, the air warms and dries. This brings periods of quite warm, dry, and stable weather to areas in the Midwest.

When the Westerlies begin to loop widely there is a greater north-to-south (that is, meridional) flow, even though the net direction of flow is from west to east. This flow causes exchange of warm and cold air. Warm air is carried northward, and cold air is carried southward. Figure 1.9 illustrates a situation in which the meandering is large enough that

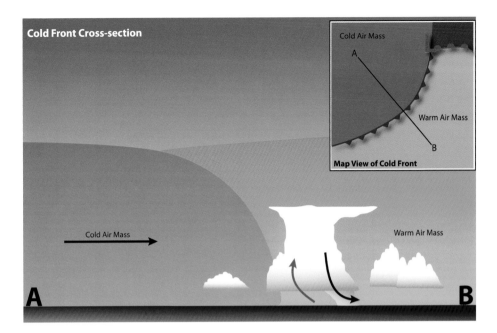

FIGURE 1.7. A schematic diagram of a cold front. The steep angle of the front results in highly unstable air at the front. The inset (upper right) shows a view of how the front appears on a weather map. Figure courtesy of Cameron Craig.

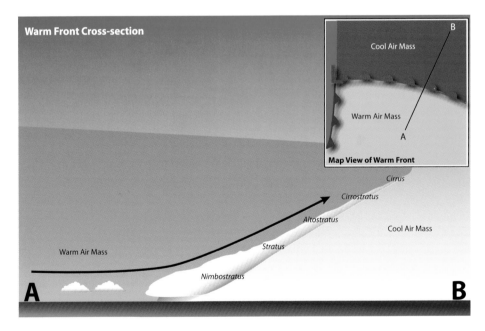

FIGURE 1.8. A schematic diagram of warm front. The low angle of the front leads to stratus-type clouds over wide areas. The inset (upper right) shows a view of how the front appears on a weather map. Figure courtesy of Cameron Craig.

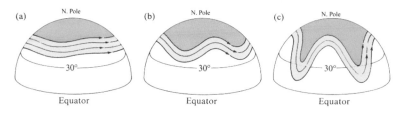

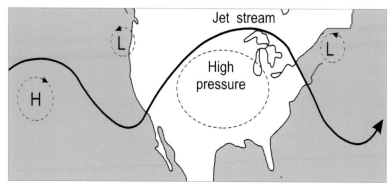

FIGURE 1.9A *(upper panel).* A highly schematic diagram of the northern hemisphere showing the development of waves in the upper air Westerlies: (a) the west to east (meridional) flow; (b) initiation of waves between the cold and warm air; and, (c) the waves deepen and this eventually lead to masses of cold air in lower latitudes and warm air in cool northern locations. The original meridional circulation is then reestablished. A jet stream is located at the leading edge of the wave.

FIGURE 1.9B *(lower panel).* A typical jet stream path across the United States is shown to be related to the location and movement of high and low pressure systems.

cells of cold or warm air masses become isolated from the main air flow. Sometimes these large masses of air stagnate and control the weather of the area in which they are located for up to several weeks; they are know as blocking systems. These masses of air can bring either very warm or very cold conditions. Blocking highs divert moisture-bringing storms away from the region and increase the probability that a drought will occur. The Westerlies shift back and forth between a zonal pattern and a meridional pattern at irregular intervals which makes long range forecasting difficult.

Jet Streams and the Seasons

Embedded in the Westerlies are cores of high speed winds called jet streams. A jet stream is a narrow band of strong winds, the speed of which varies over time and space. However, speeds of 100 to 200 mph occur frequently in given portions of the streaming air while much higher speeds are found in jet streaks.

There are two main jet streams that influence the United States. The subtropical jet occurs in the upper air at the transition between tropical and mid-latitude air. The polar-front jet is found at the leading edge of the troughs and ridges of the meandering Westerlies. The polar-front jet is of extreme importance in deciphering and predicting Indiana's weather for it is indicative of surface contrasts in temperature. The location of the polar-front jet is associated with the formation of fronts and middle latitude cyclones for which the jets act as steering mechanisms. They are, accordingly, of greatest interest to meteorologists forecasting the weather.

Jet stream locations are determined by temperature gradients. In the summer, the difference in temperature between the tropics and the poles is relatively small, and the polar jet stream typically lies sufficiently north of Indiana as to have minimal impact upon its weather. As the temperature of polar realms and the northern part of the continent becomes colder, so the jet stream and polar front migrate southwards, so that by midwinter it is located to the south of Indiana and the Midwest. By spring, the northward retreat begins so that the cycle is complete and Indiana is again influenced mostly by tropical air masses. Figure 1.10 provides a summary of these events and relates them to prevailing weather conditions.

This chapter provides the background to many of the subjects discussed in ensuing chapters. The significance of earth-sun relations cannot be overemphasized, for eventually all atmospheric events on earth relate back to energy derived from our sun. This effect may be observed in such things as daylight hours, but less easily seen in many other atmospheric features. Precipitation processes operate using energy derived from the sun, as do the circulation of the atmosphere and the winds that result.

It is impossible to understand the weather and climate of Indiana without some notion of the impact of the upper air Westerlies. Zonal flows provide very different weather from that associated with meridional flows. The former provide fairly placid conditions, the latter often give rise to outbursts of polar air or storm systems. As part of the westerly circulation, jet streams provide information about air flow and storm formation. Many instances of this will be seen in the pages that follow.

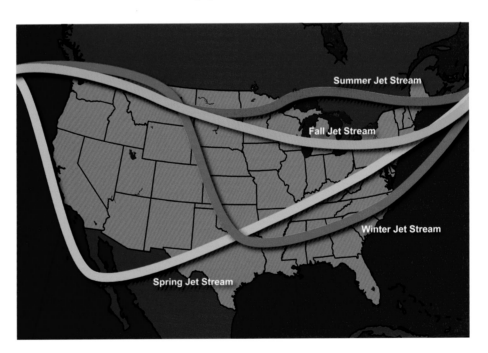

FIGURE 1.10. Over the course of a year the average location of the jet stream changes in relation to the seasonal temperature differences between low and high latitudes. Figure courtesy of Cameron Craig.

2

SPRING

Spring is perhaps the most welcomed of all seasons. The short days and snow and ice of winter gradually give way to brighter, warmer conditions and spring flowers begin to emerge. The gradual warming during spring results from the lengthening of days and the increasing intensity of the sun's rays as the sun gets higher in the sky. As warm air moves in from the south, colder air is retreating to the north and the jet stream migrates northward toward its average summer position (figure 2.1; see figure 1.10). The retreating cold air is still very cold, and the invading air is quite warm in contrast. This sets up some special conditions that can lead to very severe weather. Thunderstorms and tornadoes may result.

But the arrival of spring in Indiana is not the same throughout the state. Consider, for example, the average dates of the last winter frost in figure 2.2. By early April frost is long gone in Evansville, while residents of Fort Wayne must wait several weeks before planting any frost-prone flowers.

Thunderstorms

It must first be noted that thunderstorms occur during all seasons, but it is in spring that they become most active and lead to other severe weather events. The formation of a cumulonimbus cloud is the sign that a thunderstorm is underway. A characteristic feature of the thunderstorm, viewed from a distance, is its cylindrical shape capped by an anvil-shaped top; the anvil is caused by air diverging from the upper portions of the storm's updraft (figure 2.3).

There are a number of ways that thunderstorms form and they are named accordingly. There are three principal types of thunderstorms: the air mass (or ordinary) thunderstorm, the multicell (frontal or squall line) thunderstorm, and the supercell thunderstorm. Thunder, lightning, and hail are integral parts of a thunderstorm. While these occur during all seasons, they are discussed in chapter 3, for summer thunderstorms are an integral part of Indiana's weather pattern. The emphasis for spring will be the tornado.

AVERAGE LAST FROST DATE
SPRING FROST: 1974 - 2003)

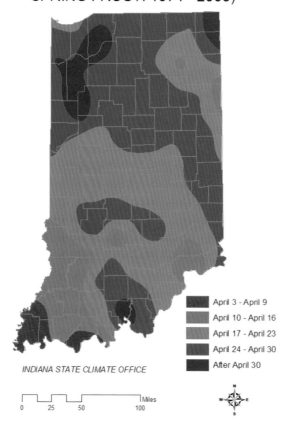

April 3 - April 9
April 10 - April 16
April 17 - April 23
April 24 - April 30
After April 30

INDIANA STATE CLIMATE OFFICE

Miles
0   25   50   100

FIGURE 2.1. *(above)*
The average location of the spring (March–May) jet stream based upon fifty years of data from the National Oceanic and Atmospheric Administration's Earth System Research Laboratory shows that it crosses Indiana. Figure courtesy of Cameron Craig.

FIGURE 2.2. *(left)*
The average date of the last spring frost in Indiana, 1974–2003. Figure courtesy of the Indiana State Climate Office.

FIGURE 2.3. A huge cumulonimbus cloud with its typical anvil shape.
Photo courtesy of Cameron Craig.

Air Mass (Ordinary) and Multicell Thunderstorms

The air mass thunderstorm is one that develops owing to solar heating. It is called an air mass thunderstorm because it forms within a uniform air mass far from fronts or other large-scale lifting mechanisms. It is the type of storm when, on a hot afternoon, the sky clouds over, still air begins to move, and a cool breeze causes leaves to be turned over. Soon a heavy downpour may occur, but generally it is over quickly.

During the roughly one hour of the storm's existence, it passes through three distinct stages: cumulus (growing) stage, mature stage, and dissipation stage (figure 2.4). In the cumulus stage, water vapor in the rising updraft condenses on microscopic particles to form cloud droplets. As condensation forms, latent heat is released which enhances the rising of the air. Cloud droplets collide and coalesce with each other to form raindrops; these raindrops continue to grow by colliding with cloud droplets and other raindrops within the updraft.

At temperatures below freezing, water vapor within the updraft is converted directly to ice crystals through the deposition of water vapor on microscopic particles. Some,

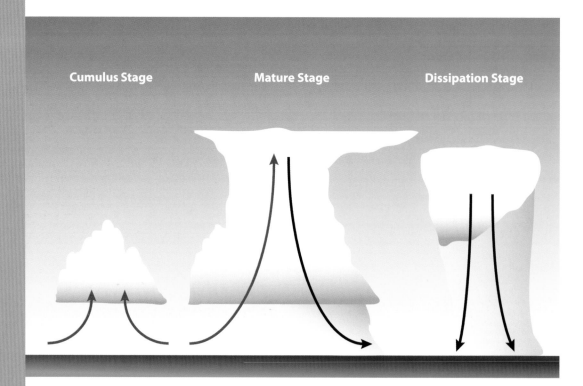

Cumulus Stage   **Mature Stage**   **Dissipation Stage**

FIGURE 2.4. Thunderstorms pass through a sequence starting with the cumulus stage in which clouds begin to grow vertically. In the mature stage the huge clouds are topped by an anvil shape caused by upper air winds. Severe weather occurs in this stage. The dissipation stage contains a downflow of cool air and the end of the storm. Figure courtesy of Cameron Craig.

however, remain in liquid form down to temperatures of -4°F (-20°C) or lower before they freeze. These are called supercooled water droplets, for they are still water when they should be ice. As discussed in chapter 3, supercooled water droplets play an important role in the formation of hail and lightning.

The beginning of the mature stage is marked by vigorous updrafts and the start of precipitation particles (rain and hail) falling toward the ground. The descending precipitation particles are accompanied by a strong downdraft of air. When the cooled air reaches the surface, it diverges outward producing a strong expanding gust front at the surface. As the surface gust front expands, it cuts off the supply of moist air to the updraft. Without an updraft to produce more precipitation and the release of latent heat, the storm enters the dissipation stage. The downdraft weakens as precipitation decreases and the storm dies.

In addition to the single thunderstorm cell, thunderstorms also organize themselves into a group of cells where each one is at a different developmental stage at a specific instant. As one cell dies so another is born. This multicell thunderstorm can become severe and produce surface damage due to strong winds and hail.

With new cells developing at periodic intervals on the right flank of the existing cells and old cells dying on the left flank, the multicell storm as a whole moves at an angle to the right of individual cell motion. This process can last for many hours, even though the lifetime of individual cells is about one hour. It accounts for those times when we think the storms will never end!

Supercell Thunderstorms

Severe thunderstorms that are capable of producing tornadoes develop an extraordinary structure during the mature stage. The supercell thunderstorm typically is an isolated storm that develops in an environment marked by extreme changing wind direction (wind shear) with height. This wind variation with height enables precipitation downdraft to form next to, and coexist with, the updraft. This feature permits the storm to remain in a mature stage for hours.

A notable characteristic of the supercell storm is its strong, rotating (spiraling) updraft (figure 2.5); the rotating column is called the thunderstorm mesocyclone. Doppler radar measurements, which show precipitation motion toward or away from the radar, indicate that the mesocyclone is typically about six miles (10 km) in diameter and extends vertically throughout much of the storm's height. If the storm produces a tornado, the tornado usually forms within the mesocyclone.

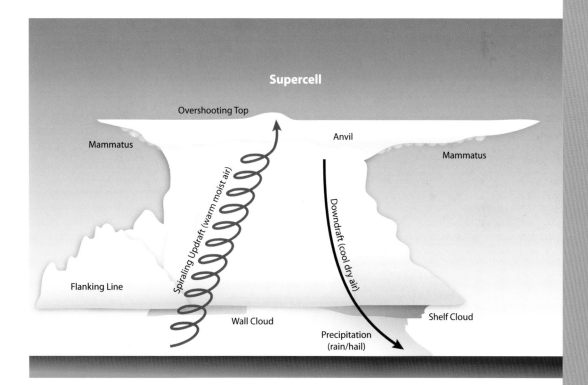

FIGURE 2.5. The anatomy of a supercell storm. Figure courtesy of Cameron Craig.

Tornadoes: Formation and Characteristics

The tornado is the most violent of atmospheric storms, but it is seldom larger than a quarter of a mile in diameter. It is a converging spiral of air with wind speeds estimated at more than 100 mph which rotates, except in rare instances, counterclockwise. The tornado begins as a funnel cloud which forms out of other clouds and works downward toward the ground; it is classified as a tornado only if it reaches the ground. The funnel cloud of condensed moisture hanging from the storm cloud is often gray in color due to condensed water vapor, but as the tip contacts the ground the appearance changes because of the dirt and debris that is picked up from the surface (figure 2.6).

Movement of the funnels and tornadoes over the ground surface is erratic, although their overall motion is dictated by the parent thunderstorm. While they normally move parallel to cold fronts, they occasionally move in circles and figure eights and may even stay in one spot. Their speed over the ground ranges from nearly stationary to as much as 65 mph (110 km/hr). The average is between 25 and 40 mph (40–65 km/hr). The surface path of most tornadoes is short and narrow, varying from a few yards to a little over one mile (2 km) in width. The path averages less than 25 miles (40 km) in length. They stay on the ground an average of fifteen to twenty minutes. Figure 2.7 shows the paths, and relative strengths, of tornadoes that occurred in Indiana between 1950 and 2001.

In order for tornadoes to form, several prerequisites are essential: (1) there must be a mass of very warm, moist air present at the surface; (2) there must be an unstable vertical temperature structure; and, (3) there must be a mechanism present to start rotation.

FIGURE 2.6. Two major tornado outbreaks have taken place on Palm Sundays. This photo of twin tornadoes was taken in Elkhart, Indiana, during the Palm Sunday outbreak in 1965. Indiana was one of six Midwest states to be raked by deadly tornadoes. In all, 47 tornadoes killed 271 people and injured over 1,500. The second Palm Sunday outbreak happened on March 27, 1994. Photo courtesy of the National Oceanic and Atmospheric Administration (NOAA); photo by Paul Huffman, *Elkhart Truth.*

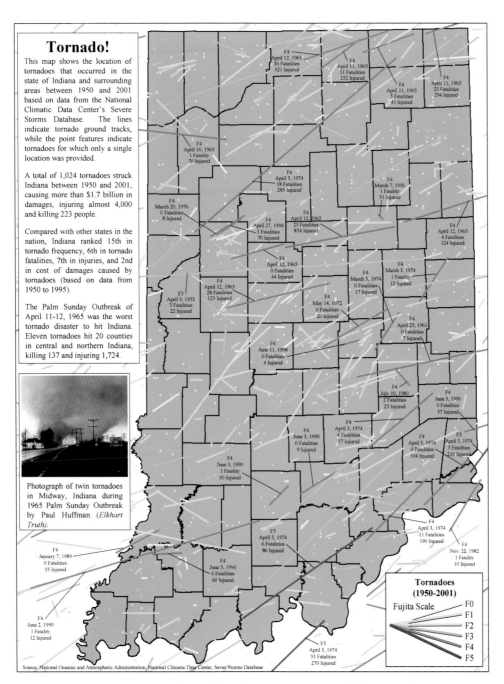

FIGURE 2.7. Indiana tornadoes and tornado tracks. Figure courtesy of the Geography Educators' Network of Indiana (GENI).

The Great Plains is the foremost tornado region in the world, and in this region a set of weather conditions often provides these three elements. Low pressure centers develop east of the Rocky Mountains. They typically have a cold front extending to the south and a warm front extending eastward from the center of the low pressure. In the warm sector there is a south-to-north flow of warm, moist air from the Gulf of Mexico. Above this air is a stream of cold, dry air from the west. This air comes from the Pacific Ocean as cool, moist air. As it crosses the western mountain ranges it loses its moisture.

The boundary between the warm, moist air from the Gulf of Mexico and the dry air from the west is a zone of great turbulence called the dry line. If the air pouring over the Rocky Mountains is warm enough, it will move out over the warm, moist air from the Gulf. The result is that the dry line extends more horizontally than vertically (figure 2.8a). This creates a stable, but potentially explosive condition. If there is some disturbance, either from the jet stream above, or from the surface below, there may be a violent exchange of air.

One such disturbance is heat from the ground surface. As the ground heats during the daylight hours, it steadily radiates more and more heat to the air above. The warm, moist air moving northward heats during the daylight hours, and by late afternoon, the surface air gets quite hot. The air near the surface becomes hot enough to break through the dry line above. This results in the explosive development of thunderstorms. The upward rush of air reaches velocities as high as 100 mph (165 km/hr). Under favorable conditions the storms will grow through the tropopause—the boundary between the troposphere and stratosphere—to heights of 60,000 feet (18,000 m). These supercells produce heavy rainfall and large hail. Violent updrafts develop in large thunderstorms and are one reason why commercial aircraft try to avoid flying through them. In order for a tornado to develop, something must start the column rotating. Wind shear can do exactly that. Wind shear is a change in wind speed and direction with height. The upper-level winds blow across the path of the lower-level winds, and at higher speeds. It is the wind shear aloft which starts the rising column of air rotating counterclockwise. As more air flows in and the storm stretches in height, the rate of rotation increases (figure 2.8b).

Tornado Intensity: The F-Scale

Tornado intensity is commonly specified according to the F-scale developed by meteorologist Tetsuya "Ted" Fujita (Fujita 1981). The scale extends from 0 to 12, but all observed tornadoes are within the lowest six classes listed in Table 2.1. F0 and F1 tornadoes are referred to as weak tornadoes, F2 and F3 storms are strong tornadoes, and F4 and F5 events are violent tornadoes. Although Fujita based his scale on damage that results from the tornado, he made efforts to approximate the wind speed ranges likely associated with each class of tornado. The actual highest measured wind speed in a tornado was 318 mph and it was measured using radar technology.

Over the years, a number of weaknesses in the original F-scale have been revealed. For example, it is based solely on the damage caused, and does not account for variations in building construction, or the absence of buildings in the tornado path. For these reason an updated version of the original F-scale was implemented in the United States on February 1, 2007. Table 2.1 presents a summary of the original and enhanced F-scales. Herein, the tornado intensities will be reported according to the original Fujita scale.

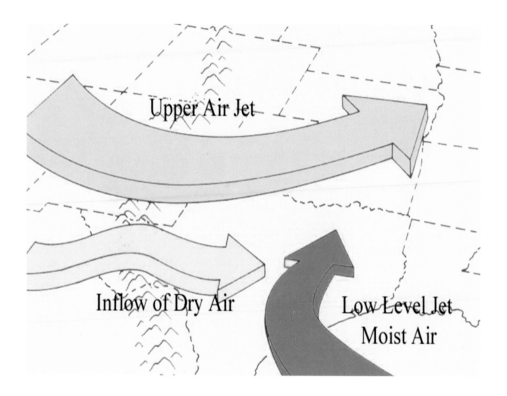

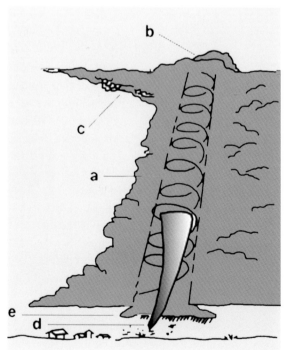

FIGURE 2.8A. *(above)*
Winds at different levels create an unstable atmosphere that gives rise to severe thunderstorms with the likelihood of tornadoes. Plentiful moisture is available at low levels which contrasts with the influx of air to form the dry line. This pattern may initially occur in the Great Plains as shown here. Figure adapted from Ahrens (1982).

FIGURE 2.8B. *(left)*
A tornado showing: (a) the mesocyclone; (b) overshooting top of cloud; (c) the anvil of the cumulonimbus with mammatus clouds; (d) the tornado; and, (e) the tornado wall cloud. Figure adapted from Ahrens (1982).

Table 2.1. Tornado Intensity Scales

Fujita (F) scale rating	Wind speed estimated from Fujita (mph)	Typical/example damage	Enhanced F-scale rating	Three-second wind gust (mph)	Total number observed in Indiana (1950–2005)
0	< 73	Damage to chimneys; branches broken off trees	0	65–85	325
1	73–112	Mobile homes pushed off foundations or overturned; moving cars blown off roads	1	86–110	389
2	113–157	Roofs torn off frame houses; mobile homes demolished; large trees snapped or uprooted; cars lifted off ground	2	111–135	261
3	158–206	Roofs and some walls torn off well-constructed houses; trains overturned	3	136–165	91
4	207–260	Well-constructed houses leveled; cars thrown and large missiles generated	4	166–200	36
5	261–318	Strong frame houses leveled off foundations and swept away	5	> 200	5

Sources: This table is derived from information provided on the following website: http://www.spc.noaa.gov/efscale/. The tornado reports shown in this table and described herein were derived from the Storm Prediction Center tornado database that includes data from 1950 to 2005. Table courtesy of S. C. Pryor.

Indiana Tornadoes

S. C. Pryor

The United States is the country with the highest frequency of occurrence of tornadoes experiencing an average of nearly nine hundred each year—but tornadoes have been observed on every continent except Antarctica. Within the United States, the state that experiences the highest number of tornadoes per square mile is Florida, followed by Oklahoma, Kansas, and then Indiana. Indiana ranks sixth in the United States in terms of the annual average number of tornado-related fatalities. In the United States between 1950 and 2005 approximately ninety-eight people died annually from tornadoes, but the death rate has declined such that over the last decade approximately sixty people are killed each year. The decline in tornado-related deaths appears to be principally linked to increased public awareness, improvements in building construction, and implementation and improvement of our warning systems. For comparison, an average of eighty people are killed by lightning each year.

Tornado Forecasting: Watches and Warnings

Tornadoes are difficult to forecast. They do not last very long, and are comparatively small in terms of atmospheric phenomena. Additionally, uncertainties remain regarding precisely how and why they form. Nevertheless, historical records have allowed meteorologists to identify atmospheric conditions that are associated with tornado occurrence. Those statistical associations are used to make a forecast of possible activity on timescales of hours to even a day or two based principally on the vertical structure of the atmosphere. If the atmosphere is forecast to have warm, moist air close to the ground surface overlain by colder, drier air, it is prone to deep mixing and possibly thunderstorm development. If this is coupled with wind shear (changing wind direction and/or speed with height) then rotation is present in the atmosphere that can be focused to generate a rotating thunderstorm and possibly a tornado. If these conditions exist, or are forecast to exist within a few hours, a tornado watch may be issued for the affected area. A tornado watch does not mean tornadoes are imminent, but rather that they are possible. A typical tornado watch covers an area of about 25,000 square miles, or over half the size of Indiana. The Storm Prediction Center of the National Weather Service issues approximately five hundred of these tornado watches each year across the United States.

When a tornado watch is issued, local storm spotter networks are activated and National Weather Service offices will begin to more closely examine data from their Doppler radar systems, along with other meteorological systems and forecast models. Doppler radar measures two key aspects of thunderstorms: first, it measures how many droplets are present in the cloud, or falling from the cloud as precipitation, based on the amount of radio waves emitted by the radar that are reflected back to the radar; and, second, it can measure the movement of the droplets using the change in frequency of the radio waves after they have interacted with the droplets. If the thunderstorm exhibits strong, highly focused rotation (as shown by the Doppler Effect), or what is referred to as a hook echo in the base reflectivity image, or if a storm spotter sees a tornado or the thunderstorm cloud evolution that typically proceeds a tornado, then a tornado warning may be issued. Under those circumstances, members of the public within that watch area are instructed to immediately take shelter.

Introduction of Doppler radars by the National Weather Service in the 1990s has greatly improved the quality of tornado

OF INTEREST
Doppler Radar

In 1842 the Austrian scientist Christian Doppler described a phenomenon that showed how the frequency of light and sound waves was affected by the relative motion between the source of the waves and the object it was detecting. This became known as the Doppler Effect.

The most frequently cited demonstration of the Doppler Effect is the change in sound waves of a train whistle as the train approaches and passes. The sound of the whistle—in particular, its pitch—becomes higher as the train approaches and lower as it moves away. This occurs because the number of sounds waves reaching an ear in a given time (the frequency) determines the pitch or tone of the sound. As a train moves closer the number of sound waves reaching an ear increases, so the pitch increases. As the train moves away the opposite occurs.

This principle is applied to bursts of extremely short radio waves, called pulses, that are transmitted from a radar antenna. By recording the direction that the antenna was facing, the location of a target is known. When the waves strike a target (which might be cloud droplets, raindrops, hail, and so on), the radio waves are reflected back to the antenna. As in the case of normal radar, the size of the target determines the strength of the reflected radio waves, while its distance from the antenna determines the time it takes for the echo to return. In this way, the location and relative sizes of cloud droplets, hail, and so on may be differentiated.

Doppler radar, however, does more than locate and determine the size of targets, it also measures information regarding their movement. The radar systems can measure the "shift in phase" between the pulse and the received echo and calculate whether the target is moving toward or away from the radar. This is enormously valuable in many ways, not the least of which is the identification of possible severe weather. If the Doppler image shows movement toward the antenna and, close by, movement away, then it is concluded that rotation must be taking place. This is exactly the movement found in a supercell.

Storm Chasing

Greg Bierly

Severe storms sweep across the Midwest (including Indiana) and particularly the Great Plains of the United States each year during the spring season. Dangerous thunderstorms develop as the powerful winds of traveling low pressure systems, leftover elements of the powerful winter circulation, meet the increasingly unstable air that has warmed over the Gulf of Mexico. The result is a meteorologist's ideal laboratory, a season and a region where high winds, hail, and tornadoes return again and again and can become the subject of field observation and measurement.

Storm interception, or "chasing," has increased in frequency (and popularity) since the 1970s, when a small group of scientists from the National Severe Storms Laboratory in Oklahoma first attempted to directly sample the atmosphere in the proximity of tornadic supercell thunderstorms with a portable instrument package. Since that time, the sophistication and the diversity of motivations behind these storm chasers has increased significantly. Storm chasers now employ a wide array of field meteorological equipment, global positioning systems, and portable Doppler radar sensors to gather data. In addition, chasers rely upon a wealth of real-time data and forecast models, much of it available via the internet, to improve their intercept capability. In particular, the advent of cell phones and wireless internet has dramatically enhanced the ability of storm chasers, professional and amateur alike, to coordinate their activity and to receive frequent, timely data transmissions. Thousands of storms chasers engage in this activity, for reasons as varied as professional research, field-based teaching, photography, personal curiosity, or simply the adrenaline rush that comes with observing powerful and elusive aspects of nature. Indeed, an entire profession has arisen in response to the demand for storm

warnings. Based on analyses conducted by the National Weather Service the percentage of tornadoes for which a warning was issued increased from 35 percent to 60 percent, and the average lead time on warnings (that is, the time between the warning being issued and the tornado touching down) increased from 5.3 to 9.5 minutes, thereby increasing the time that the public had to take shelter. These improvements are even more marked for the violent tornadoes and have decreased the number of injuries and deaths (Simmons and Sutter 2005).

Tornado Occurrence in Indiana

Between 1950 and 2005 1149 tornadoes struck the state of Indiana. Of the 1007 that have F-scale ratings, 65 percent are classified as weak tornadoes (F0 or F1), 32 percent are strong (F2 or F3), and thus approximately 3 percent are violent (F4 or F5). (Refer to Table 2.1.) Similar analyses for the entire United States indicate that approximately 68 percent of all recorded tornadoes are categorized as F0 or F1, while less than 1 percent are rated as F4 or F5. However, the 1 percent of tornadoes

that are classified as violent account for 26 percent of tornado deaths across the United States.

In Indiana tornadoes were observed on 510 days in the fifty-six-year record, giving an average frequency of tornado days in the state of 2.5 percent. This means that a tornado would be sighted somewhere in Indiana on about eight or nine days in an average year. For comparison, thunderstorms are observed in Indiana on average between forty and sixty days per year.

Where Do Tornadoes Occur?

Figure 2.10 shows the touchdown points (by F-scale rating) for all tornadoes that impacted the state of Indiana and were assigned an F-scale rating. As shown, all counties within the state experienced at least five tornado touchdowns between 1950 and 2005. There is some clustering of tornado reports

FIGURE 2.9. One of the student storm chasers from Indiana State University views a gust front associated with a cumulonimbus cloud. Photo courtesy of Greg Bierly.

chasing experiences—that of the storm chase guide—individuals who are paid to provide clients up-close experiences with tornadoes.

Although the variable topography and tree line of Indiana make it an extremely challenging (and dangerous) environment in which to intercept severe weather, several Indiana-based groups, including university teams from Ball State, Indiana State, Purdue, and Valparaiso, now routinely engage in storm intercept projects as part of either research or teaching programs. These groups typically travel in search of severe weather, traversing thousands of miles of the central United States over multi-week classes, but occasionally severe weather outbreaks occur within or near Indiana state boundaries.

Regardless of the location and objectives of the chase, or the background of the participants, storm chasing involves a balance between forecasting skill, logistics, and decision making, on one hand, and safety, on the other. While predicting a general region for an outbreak of severe weather is quite difficult (but practiced with admirable skill and precision by the forecasters of the Storm Prediction Center), determining the precise location of the initiation of convection is even harder. Often, multiple storms become active, but only selected cells become severe; anticipating which storms will develop rotation from those which may not is an even finer distinction, requiring patience, extremely careful observation, experience, and a considerable amount of luck. Finally, even if the storm chase team has singled out the correct storm to observe and pursue, it is extremely difficult to navigate often unknown roads while simultaneously following a fast-moving cell and maintaining a safe position and distance.

Storm chasing is likely to sustain the interest of meteorologists and the public for the foreseeable future. Outside of computer simulation, tornado dynamics can only be sampled and studied with the benefit of field interception. And that practice will always involve the thrill of the chase.

around the major urban areas where the greater population increases the chance that tornadoes will be observed and reported. However, counties in the northwest of the state (south of the lake counties) have a higher incidence of tornadoes even accounting for reporting bias. These counties represent an eastward extension of a region of high tornado frequency in Illinois (Wilson and Changnon 1971).

Occasionally people will ask "I've lived in Indiana all my life and I've never experienced a tornado despite hearing many warnings—how come?" As discussed above, most thunderstorms never spawn tornadoes, and even when tornadoes do occur they are typically not very long lived. Based on analyses of the path length and width reports for each tornado observed in Indiana over the last fifty-six years, the likelihood of any particular location in Indiana being affected by a tornado of any magnitude in a given year is approximately 1 in 3,000. An alternative way to think about this question is to consider what happens during tornado warnings. The false alarm rate for tornado warnings is about 76 percent. That means that only one in four warnings is associated with an actual tornado occurrence. Even when a tornado is forecast and does occur, the chance that any individual within that watch box will experience it is small. The typical size of a tornado watch box is between 1,000 and 10,000 square miles. Given that a tornado is likely to be an F0 or F1 tornado on average it will likely affect an area of less than 0.2 square miles. So assuming your house covers an area of 2,000 square feet the chance of the correctly forecasted tornado hitting your house is less than 4 in 1,000, and probably much lower than that.

Nevertheless, the low probability of actually being struck by a tornado does not mean that the warnings should be ignored. On average about seven people are killed in Indiana each year by tornadoes, and the chances of surviving a tornado strike are massively increased by seeking appropriate shelter.

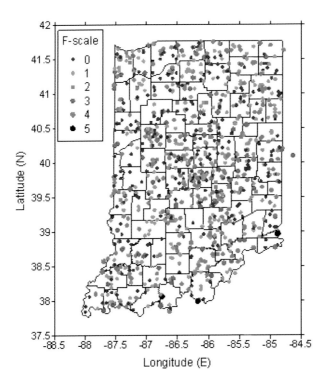

FIGURE 2.10. Spatial distribution of tornado occurrence in Indiana (1950–1995) shown by the recorded initial tornado touchdown and F-scale rating. Figure courtesy of by S. C. Pryor.

When Do Tornadoes Occur?

The occurrence of tornadoes in Indiana varies with year, month, and hour of the day (figure 2.11). Indiana experiences a high degree of year-to-year variability in tornado occurrence, with as many as fifty tornadoes in some years and as few as two in others. This inter-annual variability is even more marked for the violent tornadoes. All five of the F5 tornadoes that struck Indiana between 1950 and 2005 were observed during 1974 in a single tornado outbreak. Compared to today, the low number of tornado reports in the early 1950s may be related to changes in public awareness. The Weather Bureau started releasing tornado forecasts in 1952 and three tornadoes occurred in 1953 which were associated with approximately one hundred fatalities. Recent increases in tornado reports appear to be primarily the result of improved automated tornado detection and ease of telecommunications, since the population of the state has changed very little over the period from 1950 to 2005.

Tornadoes have been observed in Indiana during every calendar month, but the highest number of tornadoes occurs in the spring and summer—particularly during April, May, and June. June is associated with the highest number of events, but April exhibit the greatest proportion of strong and violent tornadoes. This seasonality in tornado occurrence is apparent in data from across the United States east of the Rocky Mountains and is principally due to seasonal shifts in the jet stream location and the intrusion of moist air from the Gulf of Mexico.

Tornadoes are also observed at all hours of the day, but are most frequently observed during the early afternoon and evening. This is particularly true for the strong and violent tornadoes, and follows the diurnal pattern of thunderstorms in the state. Perhaps unsurprisingly tornadoes at night tend to be associated with more deaths and injuries than those that occur during the day. They are less

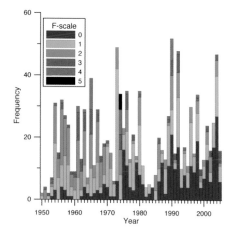

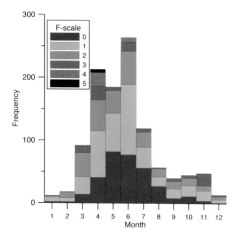

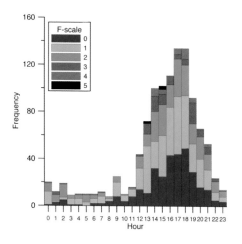

FIGURE 2.11. Tornado occurrence in Indiana (1950–1995) shown by (a) year, (b) month, and, (c) hour of the day presented as a function of F-scale rating. Figure courtesy of S. C. Pryor.

likely to be seen by the storm spotters (thus reducing the possibility of a warning being issued) and people are less likely to be aware that a warning has been issued.

Worst Tornadoes in Indiana History

The deadliest year for tornado deaths in the United States was 1925, when 794 people died. Most of these deaths (approximately 750) were associated with the tri-state tornado outbreak (an outbreak is an event in which six or more tornado touchdowns are observed in a twenty-four-hour period) that impacted Missouri, Illinois, and Indiana on March 18. The largest of the tornadoes, an F5, moved over 215 miles at forward speeds of between 60 and 73 mph.

A brief summary of the three days with highest tornado counts in Indiana (that is, the three worst tornado days) as documented in the 1950–2005 detailed data record is given in Table 2.2. April 3, 1974, is notorious as one of the worst days for severe weather in the Midwest. Five of the twenty-five tornadoes that impacted Indiana were rated as F5, and a further eight were F4. These tornadoes were part of the worst tornado outbreak in United States history. Over April 3–4, thirteen states (Alabama, Georgia, Illinois, Indiana, Kentucky, Michigan, Mississippi, North Carolina, Ohio, South Carolina, Tennessee, Virginia, and West Virginia) were impacted by 148 tornadoes. Although damage estimates and fatalities vary, in Indiana there were fifty confirmed deaths, and perhaps one thousand injured. A high likelihood for severe weather was forecast on April 2, including parts of Indiana, but the forecast area extended further to the west than was observed and did not include regions to the east that were impacted. Of the F5 tornadoes the one with the widest path (in excess of 1 mile) struck near the city of DePauw and had a damage path of about 65 miles in length. Shortly after that tornado lifted, an F4 tornado struck nearby Jefferson County destroying or damaging nearly 90 percent of the buildings in Hanover.

June 2, 1990, saw forty tornado touchdowns across Indiana, seven of which were rated as F4. A further forty-one tornadoes were observed in the surrounding states. This outbreak was associated with eight confirmed deaths in Indiana and perhaps 260 injuries. Finally, May 30, 2004, experienced twenty-four tornadoes across Indiana, two of which were F3. This event saw eighty-two tornadoes across the entire eastern United States, and in Indiana the outbreak resulted in one death and thirty-eight injuries.

Indiana experiences an average of nearly nineteen tornadoes each year which occur on approximately eight or nine days during the year. However, this statistic masks the huge year-to-year variability in tornado occurrence. Some years see very few events whereas others have over fifty days with confirmed tornado occurrences. Tornadoes have been observed in each month of the year and each hour of the day, but the largest number of the most intense tornadoes (F2 and higher) typically occur during March to June, during the afternoon and evening. Tornado warning accuracy and lead times have increased due to the introduction of Doppler radar technology, but twenty-four tornado-related deaths nonetheless occurred in Indiana during a single event in 2005.

**Table 2.2. Summary of the Three Days with the Highest Number
of Observed Tornadoes in Indiana, 1950–2005**

Date	Time (beginning–end of the event)	Total number of tornadoes in Indiana	Total number of tornadoes across the eastern U.S.A.
April 3, 1974	8 am–8 pm	25	148
June 2, 1990	6 pm–10 pm	40	81
May 30, 2004	Noon–10 pm	24	82

Source: Table courtesy of S. C. Pryor.

3

SUMMER

The lazy, hazy days of Indiana's summer are clear evidence that the fronts and storms associated with the jet stream are now typically far to the north (figure 3.1). The retreating jet stream, a result of the relatively small temperature difference between high and low latitudes, means that moist tropical air can now make its way to the state. The average July temperatures (figure 3.2) and the long hours of daylight permit outdoor activities of all kinds and this, in itself, means that dangerous weather can take its toll. Golfers, fishermen, and ballplayers, for example, can be exposed to further thunderstorm hazards.

Thunder, Lightning, and Hail

Thunderstorms are a fairly common event in summer. Given the circulation patterns and relative lack of frontal systems, the storms are often of the air mass variety. Nonetheless, the storms can be severe and severe thunderstorm watches and warnings may be issued. Recall that a watch is given when conditions are favorable for the development of severe thunderstorms in or close to an identified area. These watches are issued for large areas by the Storm Prediction Center in Norman, Oklahoma, and are usually valid for four to six hours. A severe thunderstorm warning is given after the identification of a storm that has large damaging hail (¾ inch, 20 mm, in diameter or larger), and/or damaging winds of around 60 mph (95 km/h) or greater.

The number of thunderstorms that occur in any year varies, but Indiana receives between forty and fifty each year (figure 3.3). While this is many more than occurs in Alaska, it is well below the more than one hundred experienced in some parts of Florida. In chapter 2 the hazard of tornadoes was related to thunderstorms. Even though a tornado may not form, severe storms provide other hazards, notably lightning and hail.

Lightning

Lightning results from the buildup and discharge of electrical energy between positively and negatively charged areas. Rising and descending air within a thunderstorm separates

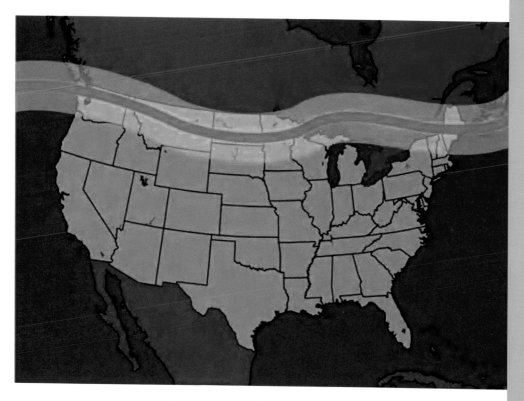

FIGURE 3.1. Average position of the summer (June–August) jet stream based upon fifty years of data from the National Oceanic and Atmospheric Administration's Earth System Research Laboratory. Figure courtesy of Cameron Craig.

FIGURE 3.2. *(below)* Average annual July temperatures in Indiana. Figure courtesy of the Indiana State Climate Office.

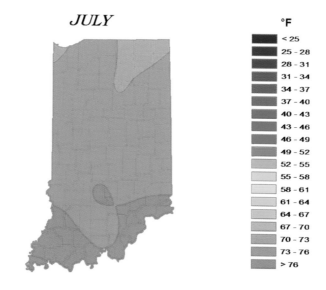

JULY

°F
< 25
25 – 28
28 – 31
31 – 34
34 – 37
37 – 40
40 – 43
43 – 46
46 – 49
49 – 52
52 – 55
55 – 58
58 – 61
61 – 64
64 – 67
67 – 70
70 – 73
73 – 76
> 76

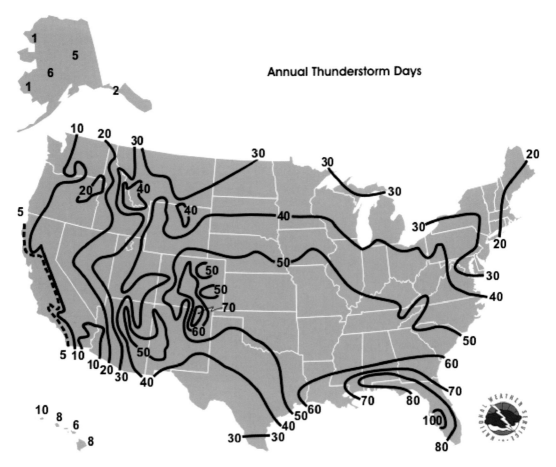

FIGURE 3.3. Average annual number of thunderstorm days in the United States. Figure courtesy of the National Oceanic and Atmospheric Administration (NOAA).

these positive and negative charges. The presence of water and ice particles plays an important part in the distribution of the electrical charges. Figure 3.4 provides some of the basic facts about lightning and illustrates the dangerous cloud-to-ground lightning stroke. This begins as an invisible channel of electrically charged air moving from the cloud toward the ground. When one channel nears an object on the ground, a powerful surge of electricity from the ground moves upward to the clouds and produces the visible lightning strike. The air near a lightning strike is heated to 50,000°F. This causes a rapid heating of the air near the lightning channel that results in a shock wave that we call thunder.

Lightning travels at the speed of light, and thunder at the speed of sound; the ratio between the two values makes it possible to estimate your distance from the lightning stroke. One need only count the number of seconds between a flash of lightning and the next clap of thunder and divide that number by five. The result provides an approximation in miles of the distance to the lightning.

Not all lightning is the cloud-to-ground form that is commonly called fork or streak lightning. Sheet lightning is seen when cloud-to-cloud lightning occurs and the channel and branches are not visible, and the cloud cover seems to light up. Heat lightning is a

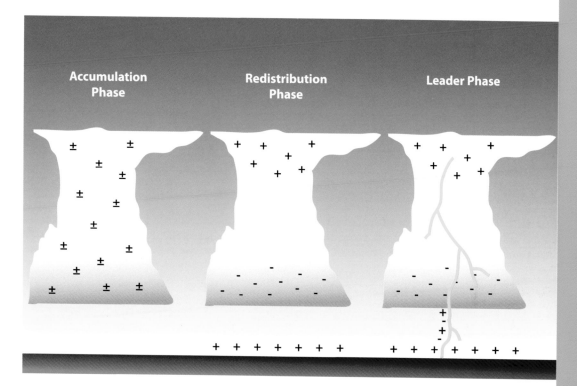

FIGURE 3.4. The separation of electrical charges in a cloud leads to lightning strikes. Figure courtesy of Cameron Craig.

misnomer. A thunderstorm some miles away lights up the distant sky near the horizon, while the local sky may be clear of clouds. This often occurs on hot, summer evenings, which is why the term "heat lightning" is used. There are other strange forms of lightning that are not often seen. Bead lightning, ribbon lightning, and ball lightning each require rather special conditions to occur.

Lightning is the most frequently encountered weather hazard, experienced by most people several times each year. In the United States, approximately one hundred people lose their lives and more than five hundred are injured by lightning every year. Most casualties result from inappropriate behavior during thunderstorms, particularly when people are caught outdoors. Thus lightning deaths and injuries are greatest in Florida where many participate in outdoor activities such as on golf courses and in sea craft. For example, in a thirty-five-year period up to the mid-1990s, 1535 deaths and injuries occurred in Florida, while in Indiana during this time period it was only 238. During the same time, 360 deaths or injuries occurred in Illinois. Only Alaska reported no injuries or deaths.

Cloud-to-ground lightning can kill and injure people by either direct or indirect means. The lightning current can be directed toward a person from tall objects such as trees and poles. The current also may travel through power or telephone lines to a person who is in contact with electric appliances, tools, electronics, or a corded telephone.

When one is struck by lightning a variety of injuries can result. The electrical current affects the many electrochemical systems in the body and people struck by lightning can suffer from severe nerve damage resulting in memory loss and personality change. The

FIGURE 3.5. Lightning photographed by Steve Sena, an American Airlines pilot. Commercial aircraft are seldom damaged by lightning. Photo courtesy of Steve Sena.

intense heat of the lightning stroke can instantly turn body moisture, such as sweat, into steam. As with the rapid production of steam from any source, the near-instantaneous expansion of high pressure steam can have dire effects including boots, shoes, and clothing being blown off a person. Burn marks may also occur in sweaty places on the body or if metal, ranging from jewelry to a belt buckle, is in contact with the body. As shown in Table 3.1 there are published rules applicable to lightning. Following the rules makes very good sense.

It is difficult to quantify losses due to lightning, especially when considering the costs of fighting forest fires caused by lightning and the cost of lightning protection to safeguard critical equipment and facilities from lightning strikes. There is little doubt that it is in the many billions of dollars annually.

Hail

Figure 3.6a shows a large hailstone cut in half. It looks like an onion, with layers of ice one upon another. This occurs because in a cumulonimbus cloud there are a number of forms of water which all play a role in hail formation (figure 3.7). Below the freezing level, water droplets occur. At higher levels, ice crystals and tiny droplets of unfrozen water are found side by side. The unfrozen droplets are called supercooled water droplets because they remain in liquid form at temperatures below the freezing point. It is this unstable supercooled water that provides the reservoir for the growth of hail. A tiny particle, consisting of salt or perhaps a microscopic pollutant, is carried in an updraft and forms an ice crystal. In its upward passage, it may come in contact with a supercooled water droplet

Table 3.1. Personal Lightning Safety Tips from the National Lightning Safety Institute

1. ***Plan*** *in advance your evacuation and safety measures.* When you first see lightning or hear thunder, activate your emergency plan. Now is the time to go to a building or a vehicle. Lightning often precedes rain, so don't wait for the rain to begin before suspending activities.

2. ***If Outdoors*** . . . *Avoid water. Avoid the high ground. Avoid open spaces.* Avoid all metal objects including electric wires, fences, machinery, motors, power tools, etc. Unsafe places include underneath canopies, small picnic or rain shelters, or near trees. Where possible, find shelter in a substantial building or in a fully enclosed metal vehicle such as a car, truck or a van with the windows completely shut. If lightning is striking nearby when you are outside, you should:

   A. *Crouch down.* Put feet together. Place hands over ears to minimize hearing damage from thunder.

   B. *Avoid proximity* (minimum of 15 ft.) to other people.

3. ***If Indoors*** . . . *Avoid water. Stay away from doors and windows. Do not use the telephone. Take off head sets.* Turn off, unplug, and stay away from appliances, computers, power tools, & TV sets. Lightning may strike exterior electric and phone lines, inducing shocks to inside equipment.

4. ***Suspend Activities*** *for 30 minutes after the last observed lightning or thunder.*

5. ***Injured Persons*** *do not carry an electrical charge and can be handled safely.* Apply First Aid procedures to a lightning victim if you are qualified to do so. Call 911 or send for help immediately.

6. ***Know Your Emergency Telephone Numbers.***

Teach this safety slogan:
"If you can see it, flee it; if you can hear it, clear it."

Source: This table is reproduced from the National Lightning Safety Institute's website and is available at http://www.lightning safety.com/nlsi_pls/lst.html.

FIGURE 3.6A. Cross-section of a large hailstone showing the onion-like structure. Photo courtesy of the National Oceanic and Atmospheric Administration (NOAA).

which freezes on it to form a thin layer of ice. As the ice moves up and down in the vigorous cloud, so other layers of ice form to give the onion-like structure. The length of time the hail stone is in the cloud ultimately determines its size, and as we know, some can be very large (figure 3.6b).

FIGURE 3.6B.
The largest hailstone to fall in the United States had a circumference of more than nineteen inches. Photo courtesy of the National Oceanic and Atmospheric Administration (NOAA).

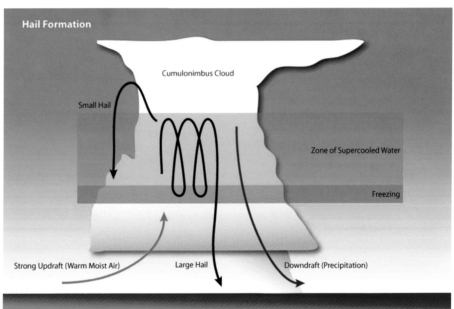

FIGURE 3.7. A schematic diagram illustrating the formation of hail. Passage of a hail stone through the zone of supercooled water leads to its layered growth. Figure courtesy of Cameron Craig.

Table 3.2. Size of Hail

Diameter	Descriptor
¼ in (6 mm)	Pea size
½ in (13 mm)	Small marble size
¾ in (20 mm)	Dime/penny/large marble size
⅞ in (22 mm)	U.S. 5 cent coin (nickel) size
1 in (25 mm)	U.S. 25 cent coin (quarter) size
1¼ in (32 mm)	U.S. 50 cent coin size
1½ in (37 mm)	Walnut or ping pong ball size
1¾ in (43 mm)	Golf ball size
2 in (50 m)	Hen egg size
2½ in (62 mm)	Tennis ball size
2¾ in (70 mm)	Baseball size
3 in (75 mm)	Teacup size
4 in (100 mm)	Grapefruit size
4½ in (115 mm)	Softball size

Newscasters often report the size of hail by likening it to common objects. There is an "official" set of descriptors for hail, and, as shown in Table 3.2, they range from pea size (a quarter of an inch diameter), to softball size (almost five inches across). The largest hail stone to be measured in the United States fell on June 22, 2003, in Aurora, Nebraska. It had a 7-inch diameter and a circumference of 18.75 inches. In Indiana, on May 5, 2000, a hailstone with a diameter of 4.50 inches was measured in Cayuga. The same size hailstone also occurred in Hartford City on April 9, 2001. The Indiana State Climate Office reports that there have been at least eight hail events where hail diameter has been greater than 4 inches.

In Indiana, hail can do tremendous damage and result in substantial economic loss for farmers. In the United States it is estimated that annual corn crop losses due to hail are some fifty-two million dollars. So significant is the loss that crop insurance guidelines to assess hail damage on most crops are available to adjusters. Hail does the most damage to corn after emergence when the growing point is well above the soil surface and plants remain vulnerable through tasseling. Unfortunately for farmers, this coincides with the period from June to September when one-third of all hailstorms occur in the United States.

Extremes 1: Floods

A flood is defined as the condition in any stream or lake when it rises above bank full, that is, the point at which the river channel is completely full with water. All natural stream floods are due primarily to surface runoff, which may result from heavy rainfall, the melting of snow, or a combination of both. Floods caused by rain can result from either short periods of high intensity rainfall or from prolonged periods of steady rains lasting for several days or weeks.

Flood runoff from small drainage basins is usually the result of different causes than floods on large drainage basins (a drainage basin is the area that supplies surface runoff to the stream or river that is draining the area). Small watersheds are on average twenty-five square kilometers or less. Watersheds of this size can be completely covered by a single convective storm, and most floods on small drainage basins are caused from cloudbursts. The rainfall is so intense that the stream channels cannot carry off the water as quickly as it falls. This leads to flash floods. On large drainage basins extended precipitation from cyclonic storms or massive snowmelt is required to produce flooding. The Wabash River has a very large drainage area and floods frequently occur in its flood plain (figure 3.8).

A series of intense thunderstorms drop enormous amounts of rain in fairly short intervals. Water that cannot enter the ground runs off into rivulets, drainage ditches, streams, and ponds. The more rain that falls, the more likely it is that flooding will occur. As the following paragraphs show, there are variations in extreme precipitation across the state and over time.

Extreme Precipitation in Indiana

J. A. Howe and S. C. Pryor

The great Midwest Floods of 1993 was one of the most widespread natural disasters in United States history and resulted in nearly fifty deaths and substantial damage to fifty-five thousand homes. Flood waters covered an area of about 325,000 square miles (840,000 sq. km), flood-related damages totaled fifteen billion dollars, hundreds of levees failed, and thousands of people were evacuated. This event, like many other floods in the Midwest, was at least partly a result of an extended period of heavy precipitation. The disaster was featured as the cover of *Time* magazine, but in many regards, particularly its impact on Indiana, it is dwarfed by the Ohio River Flood of 1937. During January 9–25, 1937, Madison, located in southern Indiana on the banks of the Ohio River, received 16.7 inches of rain. Indeed, January 15, 1937, holds the record for the fifteenth highest daily precipitation value for a twenty-four-hour period over the past century, recording 3.55 inches of rain. During the Ohio River Flood of 1937, from small towns in southern Illinois to Pittsburgh, Pennsylvania, one million people were rendered homeless, 385 people lost their lives, and five hundred million dollars worth of property was destroyed or damaged. These two events underscore why flood damage ranks as the top weather-related cause of economic loss in the United States.

Calculating Extremes

Various methods are used to examine extreme precipitation events. Percentiles, which are used here, provide the spread or variation of a large set of numbers by determining the fraction or percentage above or below given values. For example, the 90th percentile daily precipitation level at a location represents the daily precipitation total exceeded on ten out of one hundred days with precipitation—or, equivalently, the daily total that was not exceeded on ninety out of one hundred rain days. Since precipitation occurs in Indiana approximately one in three to four days, this daily precipitation amount is reached or exceeded on an average of approximately ten days per year.

Figure 3.9a shows the 90th percentile daily precipitation at ten sites across Indiana computed based on data from 1896 to 2002 compiled by the Illinois State Water Survey. As shown, the 90th percentile daily precipitation values exhibit a north-south gradient across Indiana, with the northern portion of the state characterized by 90th percentile

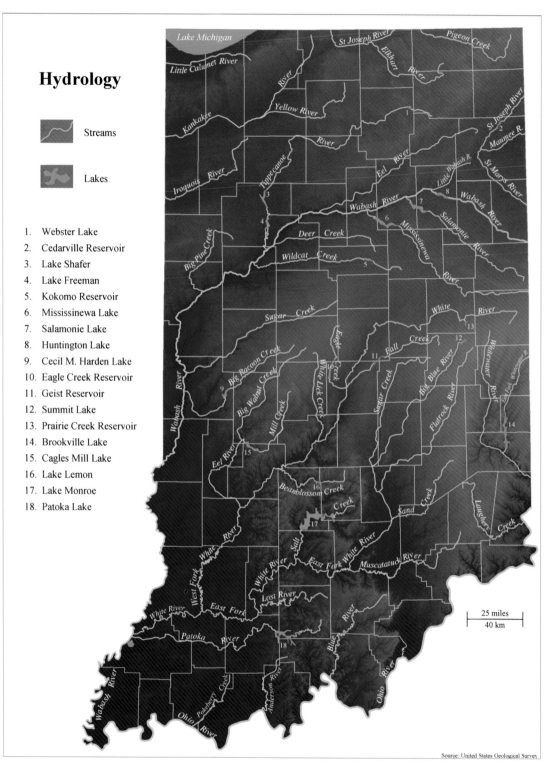

Hydrology

Streams

Lakes

1. Webster Lake
2. Cedarville Reservoir
3. Lake Shafer
4. Lake Freeman
5. Kokomo Reservoir
6. Mississinewa Lake
7. Salamonie Lake
8. Huntington Lake
9. Cecil M. Harden Lake
10. Eagle Creek Reservoir
11. Geist Reservoir
12. Summit Lake
13. Prairie Creek Reservoir
14. Brookville Lake
15. Cagles Mill Lake
16. Lake Lemon
17. Lake Monroe
18. Patoka Lake

Source: United States Geological Survey

FIGURE 3.8. The major rivers and lakes of Indiana. Figure courtesy of the United States Geological Survey and the Geography Educators' Network of Indiana (GENI).

daily precipitation totals ranging from 0.33 to 0.35 inches, while the south experiences higher values, ranging from 0.41 to 0.42 inches.

Although northern Indiana experiences lake-effect snowfall during the winter, the low ratio of liquid water to snow volume (typically, 1 inch of snow equivalent to 10 inches of water) in those events and higher total precipitable water in southern Indiana causes higher precipitation accumulations to be observed in the south during spring and summer. Indeed, for all of the sites presented, the highest daily precipitation values occur between March and September.

Of the ten Indiana sites depicted in figure 3.9a, the highest daily precipitation value for 1896–2002 occurred in Valparaiso on August 27, 1923, with 7.34 inches falling in just twenty-four hours. The second highest daily precipitation value occurred in Bloomington on March 25, 1913, with 6.56 inches of precipitation. On that same date, Washington, which is just southwest of Bloomington, also received its own record high daily precipitation at 5.83 inches.

At the national level, flood-related annual losses have increased from one billion dollars in the 1940s to six billion dollars in the 1990s (both figures adjusted to 1997 dollars). Given that floods are related to intense precipitation events, it has been postulated that part of the increase in flood-related economic losses across the United States is due to an increase in the frequency of extreme precipitation events. One study found that over the twentieth century, total annual precipitation across the United States increased by approximately 10 percent, with more than half of the increase being attributed to an increase in heavy precipitation events. A 2001 study further suggested that in some parts of the United States, such as the Midwest, annual total precipitation will further increase by between 10 and 30 percent over the next century resulting in concerns about a possible increase in flood frequency.

In light of research that has indicated the presence of temporal trends in precipitation records from the contiguous United States during the latter portion of the twentieth century, daily precipitation data at these ten sites in Indiana, along with ten sites from each of the other midwestern states, were analyzed for temporal trends using linear regression. The results (shown in figure 3.9b) indicate that sites from Indiana and other southern states of the Midwest exhibit a trend toward lower values of the annual 90th percentile precipitation over the past century. Data from stations across the northern Midwest tend to exhibit an increase in the annual 90th percentile precipitation over the last century.

Extremes 2: Drought

This section deals with meteorological drought, and uses data and information provided by Umarporn Charusombat of the Indiana State Climate Office at Purdue University.

Drought is the world's costliest natural disaster and one that is well known to Hoosiers, especially Hoosier farmers. The *Glossary of Meteorology* (1959) defines a drought as "a period of abnormally dry weather sufficiently prolonged for the lack of water to cause serious hydrologic imbalance in the affected area." Despite this seemingly simple definition, drought can be defined in other ways. Meteorological drought is a prolonged period without rain, with the length of the dry period related to the regional climate. Agricultural drought relates not only to precipitation amounts but also to evaporation and transpiration, soil water deficits, and ground water or reservoir levels. Agricultural drought also can be defined as the variable susceptibility of crops to lack of moisture during different

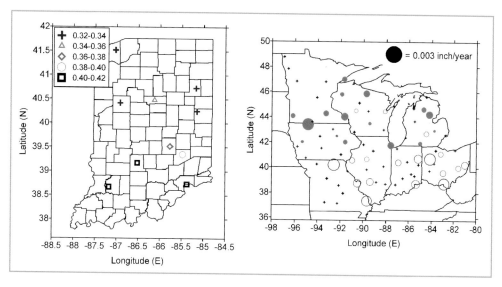

FIGURE 3.9A *(left).* Extreme (90th percentile) daily precipitation values (in inches/day) at ten sites in Indiana. The higher values experienced in the south are shown. Figure courtesy of S. C. Pryor.

FIGURE 3.9B *(right).* The magnitude of linear trends (1896–2002) in extreme (90th percentile) precipitation at eighty sites across the Midwest. If the dot is filled the trend is toward increasing values; if the dot is open the trend is toward decreasing value. If the station is denoted by "+" no trend is observed. The magnitude of the trend is shown by the size of the dot. As shown, some sites in Indiana experienced a decreasing trend in this time period. Figure courtesy of S. C. Pryor.

stages of crop development, from emergence to maturity. Hydrological drought is associated with the deficiency of water supply such as stream flow, reservoir and lake levels, and ground water. Finally, socioeconomic drought concerns the demand for and damage to any economic entity that is susceptible to meteorological, hydrological, and agricultural drought. Drought also can be estimated simply by delineating periods ranging from three to sixty months, and setting criteria for its definition. Illinois uses a scale of criteria that depends on the statewide mean precipitation departure from normal and the areal extent of the precipitation deficiency.

Worst Case Droughts in Indiana

As mentioned above, drought due to precipitation deficiency can be determined from the precipitation departure from normal. Given that the statewide annual normal precipitation for Indiana is 41.49 inches it is possible to identify drought years. The driest year on record for Indiana was in 1963 when precipitation averaged 70 percent of normal, or 29.32 inches. Table 3.3 lists the driest years on record in Indiana.

Between 1987 and 1988 many parts of the United States experienced severe drought that was shown to be related to the strength of El Niño. Indiana did not show the same relationship, since the statewide annual precipitation, 34–35 inches for that period, was 83–84 percent of normal.

Of the many drought indices, three are perhaps the most widely used, the PDSI, the Crop Moisture Index (CMI), and the Standardized Precipitation Index (SPI).

The PDSI is used by many government agencies to trigger drought relief agencies. It employs a complex formula and is used to identify and measure dry periods over many months. It allows standardized measures of moisture conditions that permit comparison by region and by month. Recent maps may be found at:

http://www.cpc.ncep.noaa.gov/products/
analysis_monitoring/regional_monitoring/
palmer.gif

The CMI is a derivative of the PDSI for use in short-term periods rather than for long-term drought and is based upon mean temperature and total precipitation for each week. It evaluates moisture levels in major crop-producing regions and responds rapidly to changing wetness conditions. It is found at:

http://www.cpc.ncep.noaa.gov/products/
analysis_monitoring/regional_monitoring/
cmi.gif

The SPI is an index based upon the probability of precipitation for a multiple time scale. Originally designed to identify the SPI for three- and forty-eight-month time scales, it can potentially provide early warning of drought and help identify relative severity. Details and monthly maps are found at the Drought Mitigation Center at the University of Nebraska, Lincoln:

http://drought.unl.edu/monitor/spi.htm

A quantitative measurement of drought is the Palmer Drought Severity Index (PDSI) which calculates levels of dryness using a water balance equation. The variables used in the equation are precipitation, temperature, evapotranspiration, runoff, and moisture loss from surface layers. The PDSI varies from 6 to -6 (wet to dry, respectively, with most values in the United States being between 4 and -4), and maps these values for current conditions. Figure 3.10 is an example of the PDSI for the United States showing the intense drought that occurred in the southeast during 2007.

An evaluation of the PDSI values since 1895 shows that Indiana's most severe drought occurred in 1964 (Table 3.4). It is interesting to note that thereafter, there is no clear correlation between the data in Table 3.4 and precipitation records. This is because many variables other than precipitation are used in preparing the PDSI.

In her study, Umarporn Charusombat uses many interesting and well-produced maps and useful graphs. They collectively provide a fine exhibit of drought in Indiana. An analysis of the maps found that most parts of Indiana have not been faced by long periods of drought. As a whole, the driest parts of Indiana are in the northern and eastern parts of the state; the wettest part is in the south.

Table 3.4. The Driest PDSI Values in Indiana, 1895–2006

Year	PDSI
1964	-3.86
1963	-3.14
1914	-2.92
1940	-2.48
1953	-2.35
2005	-2.30
1895	-2.27
1930	-2.25
1901	-2.20
1956	-1.95

Table 3.3. The Driest Years on Record in Indiana, 1897–2006

Year	Annual precipitation (inches)	% of Normal
1963	29.32	70.67
1930	29.5	71.30
1934	29.66	71.50
1901	30.58	73.70
1953	31.19	75.17
1914	31.43	75.75
1940	32.66	78.72
1941	32.85	79.18
1976	33.02	79.58
1960	33.99	81.92

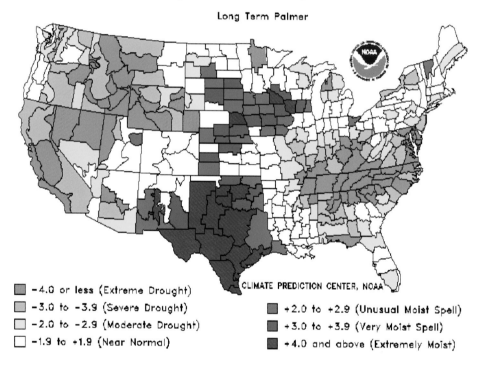

FIGURE 3.10. The Palmer Drought Severity Index, issued by the National Oceanic and Atmospheric Administration on a weekly basis, illustrates the extremely dry conditions in both eastern and western areas of the United States during 2007. Figure courtesy of the National Oceanic and Atmospheric Administration (NOAA) and the National Aeronautics and Space Administration (NASA).

People and Summer

Cooling Degree Days

Summers in Indiana can be very warm, and at times quite uncomfortable. The modern way to beat the heat is to use air conditioning. Upon entering most stores, malls, and many homes on a hot summer day one is frequently met by what seems, at first, a rather chilly blast of air. After a while, as the body adjusts to the new temperature, it becomes quite comfortable. This comfort does not come without cost, and most of the cost in seen in utility bills. To provide a guide to how much these bills might be, cooling degree days (CDDs) were devised. The idea of the degree day was formulated by heating engineers and, in the United States, is computed using the Fahrenheit scale. CDDs are generally computed based upon a base outside air temperature of 65°F, although sometimes a higher base temperature, often 70°F, is used. CDDs are computed in the following way:

> Assume the average outside temperature is 75°F for a given day
>
> With a CDD base of 65°F, then 10 CDDs (75-65) are accumulated
>
> The average on the next day may be 77°F
>
> For that day, 12 CDDs (77-65) are accumulated

The 22 CDDs for those two days are added to all the days when the temperature is above the base temperature to result in a total number for the entire summer season. Obviously, the hotter the day, the more CDDs accumulated.

Average CDDs for Indiana are shown in figure 3.11. The energy needs are the inverse of the costs of heating; the highest values occur in the southern counties and many fewer high values occur in the northern counties (see the discussion of heating degree days in chapter 5). The summer utility costs of the south are counterbalanced by the lower utility costs of winter—and vice versa for the northern counties.

Sunburn

Sunlight arriving at the top of the atmosphere contains ultraviolet, visible, and infrared radiation. Ultraviolet radiation (UV) has different effects at different wavelengths so three bands are recognized: UV-A, UV-B, and UV-C, from longest to shortest wavelengths. The highly damaging UV-C is absorbed by stratospheric ozone, while UV-A and UV-B travel to the earth's surface.

Radiation absorbed at the body surface has a number of effects. Of prime importance is the production of vitamin D, the vitamin necessary for the prevention of bone disease. Another well-known effect of ultraviolet radiation absorption in humans is sunburn. Exposure to UV radiation produces a capillary-dilating chemical which immediately induces a reddening of the skin and ultimately, with continued exposure, blistering. Differences in skin types influence the rates of reaction. A non-tanned, white-skinned person will show traces of skin reddening after three to twenty minutes of exposure on a clear Indiana midsummer day. Further exposure leads to a photochemical response of the skin and the production of melanin. Extensive production of melanin over a long period of time leads to skin that appears to age rapidly and it possibly may result in skin cancer.

Given that both short- and long-term exposure to UV radiation is a health hazard, the National Oceanic and Atmospheric Administration and the Environmental Protection Agency put into use the UV Index that predicts UV radiation levels for noon standard time

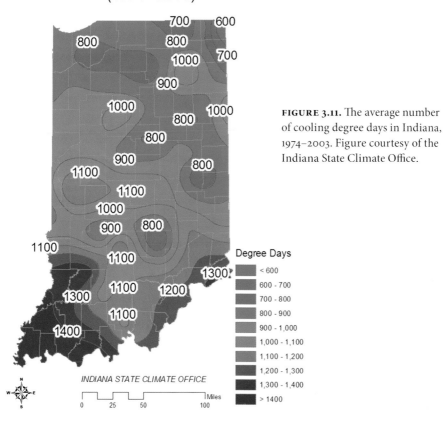

AVERAGE COOLING DEGREE DAYS
(1974 - 2003)

Degree Days

- < 600
- 600 - 700
- 700 - 800
- 800 - 900
- 900 - 1,000
- 1,000 - 1,100
- 1,100 - 1,200
- 1,200 - 1,300
- 1,300 - 1,400
- > 1400

INDIANA STATE CLIMATE OFFICE

Miles
0   25   50   100

FIGURE 3.11. The average number of cooling degree days in Indiana, 1974–2003. Figure courtesy of the Indiana State Climate Office.

for many cities including Indianapolis. The UV Index is provided with the daily forecast and is derived using satellite and ground-based observations and computer models.

Heat and Heat Waves

Although not officially defined in quantitative terms, the *Glossary of Weather and Climate* published by the American Meteorological Society states that a "heat wave is an extended period of abnormally and uncomfortably hot and usually humid weather. To be a 'heat wave' such a period should last at least one day, but conventionally it lasts from several days to several weeks" (Geer 1996). Others concur with this definition (Robinson 2001).

Indiana has had its share of heat waves. In Evansville a temperature of over 90°F was recorded 105 times in 1953. The city also suffered in 1955 when the temperature was above 100°F on thirty-three days. However, the highest recorded temperature belongs to Collegeville when the maximum temperature was 116°F in 1936. Such high temperatures can be lethal. Figure 3.12 shows the intensity of the July 1999 heat wave in the Midwest. Southern Indiana had in excess of twelve days over 90°F.

Heat kills by taxing the human body beyond its abilities. The actual number of heat-related deaths is hard to define precisely because no one can know how many additional deaths are caused by the hot weather over and above those deaths that would have occurred without the heat stress. Indiana summers are hot and the hot conditions tend to combine both high temperature and high humidity.

The body reacts to such conditions by attempting to dissipate heat in a number of ways. The heart pumps more blood to the outer layers of skin in order to circulate blood closer to the skin-air interface. This permits heat from the body to pass to the surrounding air. The red face of overexertion is but one expression of this. At the same time, sweating begins and perspiration appears at the skin's surface.

Cooling of the body by sweating is based upon the principle that when liquid changes state, such as from liquid to gas through evaporation, heat energy is absorbed. This extracts heat from the body and thus acts as a cooling mechanism. However, the rate at which the sweat evaporates depends not only on temperature, but also on how much moisture the surrounding air contains. If, for example, the relative humidity is low, the moisture will evaporate rapidly. If engaging in an activity, such as playing tennis, causes one to sweat, after one stops the activity it can feel surprisingly cool. But when humidity is high, as is often the case in Indiana's summers, the evaporation occurs much more slowly and the sweat will remain on the skin. This is the reason for those familiar sticky days when one becomes soaking wet from perspiration that has not evaporated.

It is important to act sensibly in heat wave conditions. There is a limit to the body's ability to shed heat through sweating and circulation changes. When the body's heat gains

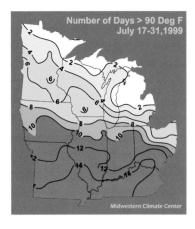

FIGURE 3.12.
Number of days over 90°F,
July 17–31, 1999. Figure courtesy of
the Indiana State Climate Office.

exceed its heat losses, or when the body cannot compensate for fluids and salt lost through perspiration, the temperature of the body's inner core begins to rise and heat-related illness may develop.

The relationship between high temperatures, high humidity, and physical comfort led the National Weather Service to devise the Heat Index (HI), a value that is often referred to as the apparent temperature. The HI, given in degrees Fahrenheit, is a measure of how hot it really feels when relative humidity (RH) is considered in relation to actual air temperature. Figure 3.13 provides information about the HI. To find the HI, look at the Heat Index Chart. As an example, if the air temperature is 95°F (found on the left side of the table) and the RH is 55 percent (found at the top of the table), the HI—or how hot it really feels—is 110°F. This is at the intersection of the 95° row and the 55 percent column. Since values were devised for shady, light wind conditions, exposure to full sunshine can increase HI values by up to 15°F. Strong, very dry winds can add appreciably to moisture loss and can be extremely hazardous.

								AIR TEMPERATURE (F)											
	80	**82**	**84**	**86**	**88**	**90**	**92**	**94**	**96**	**98**	**100**	**102**	**104**	**106**	**108**	**110**	**112**	**114**	**116**
5	78	80	81	83	85	86	88	89	91	93	94	96	97	99	101	102	104	105	107
10	78	80	81	83	84	86	88	89	91	93	95	97	99	101	103	105	107	109	111
15	78	80	81	83	84	86	88	90	92	94	96	98	100	103	105	108	111	113	116
20	79	80	81	83	85	86	88	90	93	95	97	100	103	106	109	112	115	119	122
25	79	80	82	83	85	87	89	91	94	97	100	103	106	109	113	117	121	125	129
30	79	80	82	84	86	88	90	93	96	99	102	106	110	114	118	122	127	132	137
35	80	81	83	85	87	89	92	95	98	102	106	110	114	119	123	129	134	140	146
40	80	81	83	85	88	91	94	97	101	105	109	114	119	124	130	136	142	148	
45	80	82	84	87	89	92	96	100	104	109	114	119	124	130	137	143			
50	81	83	85	88	91	95	99	103	108	113	118	124	131	137					
55	81	84	86	89	93	97	101	106	112	117	124	130	137						
60	82	84	88	91	95	100	105	110	116	123	129								
65	82	85	89	93	98	103	108	114	121	128	136								
70	83	86	90	95	100	106	112	119	126										
75	84	88	92	97	103	109	116	124											
80	84	89	94	100	106	113	121												
85	85	90	96	102	110	117													
90	86	91	98	105	113					Blank area is where dewpoint > 85F									
95	86	93	100	108															
100	87	95	103																

RELATIVE HUMIDITY (%)

Heat disorder risk   CAUTION   EXTREME CAUTION   DANGER   EXTREME DANGER

FIGURE 3.13. The Heat Index Chart provides the apparent temperature resulting from the combined effects of temperature and relative humidity on humans. The hazards posed by heat stress are shown in the legend. Figure courtesy of the National Oceanic and Atmospheric Administration (NOAA) and the National Weather Service (NWS).

4

FALL

Throughout fall the weather gradually changes, with the midday sun lower in the sky and the days growing shorter. The often-localized showery weather of summer is replaced by huge migrating weather systems. Polar regions become colder and the cold air, guided by a jet stream, advances southward to meet the moist tropical air that remains in place after the heat of summer (figure 4.1). Fronts are formed, and Indiana alternatingly experiences gloomy, rainy days and cool days with sunny skies. The change from the warmth of summer to the cold of winter is on its way.

People become aware of the cooling, especially when the first frost of the season occurs. As in the case of the last frost in spring, the date of the first frost varies from the south of the state to the north. As figure 4.2 indicates, there is quite a time difference between the average dates of fall's first frost across Indiana. Many of the northern counties experience their first frost in early October, while in the southern counties, the average date is after October 23. The south has a growing season, as measured by the average number of frost-free days, which is appreciably longer than that of the northern part of Indiana (figure 4.3).

A Quiet Transition Season

Spring is the transition from winter to summer weather. It is, as we have noted, a time of beauty that is tempered by violent weather. Fall is also a transition season, from summer to winter, but one with much less stormy weather. This difference occurs because the contrast between the temperatures of polar air and tropical air is much less in fall than it is in spring. This is to be expected because the air of the polar regions has not yet had time to become the frigid source region caused by the lack of sunlight of long winter nights. As fall progresses, deciduous trees lose their leaves and the green of summer slowly gives way to the bare, brown landscapes of winter.

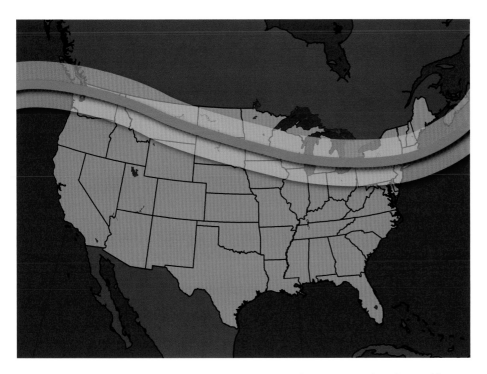

FIGURE 4.1. Average position of the fall (September–November) jet stream based upon fifty years of data from the National Oceanic and Atmospheric Administration's Earth System Research Laboratory. Figure courtesy of Cameron Craig.

AVERAGE FIRST FROST DATE
(FALL FROST: 1974 - 2003)

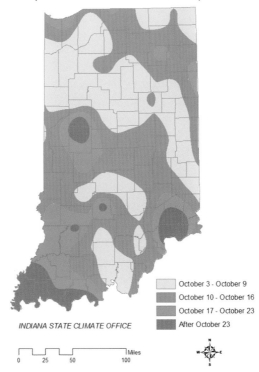

October 3 - October 9
October 10 - October 16
October 17 - October 23
After October 23

INDIANA STATE CLIMATE OFFICE

0   25   50   100   Miles

FIGURE 4.2. The average date of the first fall frost in Indiana, 1974–2003. Figure courtesy of the Indiana State Climate Office.

AVERAGE NUMBER OF FROST-FREE DAYS
(GROWING SEASON 1974 - 2003)

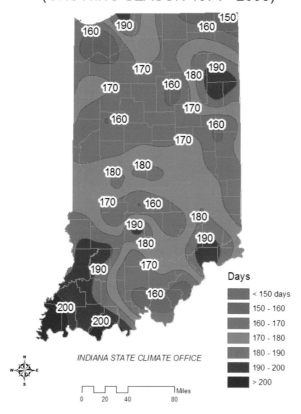

Days

	< 150 days
	150 - 160
	160 - 170
	170 - 180
	180 - 190
	190 - 200
	> 200

INDIANA STATE CLIMATE OFFICE

Miles
0   20   40   80

FIGURE 4.3. Average length of the growing season in Indiana, 1974–2003. Figure courtesy of the Indiana State Climate Office.

Autumn Leaves

The leaves of trees have been described as nature's food factories. Through photosynthesis—a word formed by the union of words that mean "light" (photo) and "joining together" (synthesis)—leaves use water obtained from roots, carbon dioxide from the air, and sunlight to produce glucose, a type of sugar. This process is enabled by chlorophyll, a chemical that gives leaves their green color. Glucose provides the basic food and energy for a tree's growth.

During Indiana's springs and summers there is a plentiful supply of each required ingredient for photosynthesis. The greenness of woods and forests are ample testimony to the productivity of the land and abundant chlorophyll. But, as summer ends and autumn arrives, days grow shorter and sunlight becomes less intense. Eventually, there is not enough sunlight and water for photosynthesis, and chlorophyll will disappear from leaves. The bright green will disappear and leaves begin to change color.

The color change occurs when the separation layer at the base of each leaf fails to transfer its manufactured glucose and energy from the leaf to the tree. Glucose and by-products remain in the leaf and, without water, chlorophyll disappears. The green color is replaced by a variety of other colors. The gold color of maples is a result of trapped glucose, while the brown of oaks results from waste products, often tannin, left in the leaves.

Over time, cells that make up the separation layer slowly disintegrate to form a tear line. The leaves separate from the tree and the bare branches of winter appear. The chemistry of the entire process of leafing, greening, changing color, and eventually separating from the tree is extremely complex and there are some parts of the process that are still not fully understood. But, irrespective of this, fall colors provide one of nature's most spectacular and unforgettable sights.

Dew, Frost, Mist, and Fog

Dew, mist, and fog can and do occur during any season. But it is in fall that we start to observe it more closely, for we know that on one morning the dew will be in its solid form, frost. On a cool, calm night, the air near the ground loses heat by radiation to the air above. With little mixing of the air, a thin layer of air next to the ground becomes cold. If the temperature drops far enough, then condensation will exceed evaporation and moisture condenses on the ground in the form of dew. The temperature at which this transition occurs is the dew point temperature. At temperatures below freezing, it is called the frost point rather than dew point. This is the temperature to which air must cool in order for ice crystals to form from water vapor. The formation of layers of ice on a car window after a cold night is a clear sign of cooling to the frost point. Often, because of the way that heat escapes from an object, such as a car, composed of different materials, ice appears on the windows and nowhere else.

Because condensation occurs at the dew point of air at a given temperature it can lead to the formation of mist and fog as well as the deposition of dew. If condensation of water droplets occurs in the layers of air immediately above the ground, it forms mist or fog. Mist is the suspension of microscopic water droplets that reduces visibility at the earth's surface. It forms a fairly thin grayish veil that

Water occurs as a liquid (water), a gas (invisible water vapor), and a solid (ice). In looking at the gaseous content of the air we center our attention upon humidity. Generally, what most people understand by humidity is actually relative humidity (RH). This is the ratio of the amount of water vapor that air contains at a given temperature to what the same air can contain when saturated. Perfectly dry air has an RH of 0 percent and totally saturated air, 100 percent. Although RH is the most commonly used measure, it does have disadvantages because the amount of moisture air can contain varies with its temperature. For example, suppose that the actual amount of moisture in the air is 4 g/kg (grams per kilogram). If the air temperature is 10°C, then the saturation value is about 8 g/kg. This gives an RH of 50 percent. But suppose that the air temperature is 20°C but it still has a vapor content of 4 g/kg. The warmer air can now hold about 16 g/km of moisture at saturation. Despite the same amount of water vapor, the RH is now only 25 percent. The same amount of moisture in the air will thus give different RH values at different temperatures.

The dew point, or dew point temperature, is another widely used measure of moisture in the atmosphere. It is the temperature at which air is saturated, that is, its RH is 100 percent. Because the amount of moisture air can contain depends upon its temperature, by lowering the temperature of an air mass, the point is reached where, because it is saturated, moisture will be deposited. This occurs at the dew point for that particular air mass.

FIGURE 4.4. Autumn leaves in Indiana. Photo courtesy of Loretta Oliver.

covers the landscape. Fog, in contrast, is the suspension of small water droplets in the air that reduces visibility to five-eighths of a mile (less than one km). In fog, the air feels raw and clammy and, given the correct illumination, the fog droplets are often visible to the naked eye. Mist does not provide the same damp, raw feeling and the individual droplets are too small to see.

While all fogs look the same, their causes are quite variable. Advection fogs form in Indiana mostly by the transport of warm air over a cold surface (advection is the horizontal transport of air; figure 4.5). Such fogs can occur when, after a cold snap, warm air moves over the cold surface. The warm air cools when in contact with the cold surface, and, if the dew point is reached, then condensation will occur in the form of fog. Such fogs are often widespread and last until the ground warms up and air temperature rises above the dew point.

Radiation fog is an extension of the way that dew forms. It occurs under clear skies in relatively still air and forms when nighttime cooling of the ground results in a chilling of

FIGURE 4.5. An advection fog can last several days and extend over large areas.

the layers of air near the surface. If the temperature drops to the dew point, then a layer of fog forms. Compared to advection fog, radiation fog lasts but a short time and "burns off" in the early morning when the sunshine heats the layer of chilled air.

A really interesting form of fog is often seen in the early morning when passing a lake or large pond. There appears to be wisps coming from the pond, and it looks almost as if the water is steaming. This occurs when the water warms a thin layer of air immediately above it, and evaporation occurs into that layer from the water surface. But the air above the thin layer is cooler, and condensation occurs, to be seen in the streams of "steam" coming from the water.

Air Quality

S. C. Pryor and A. M. Spaulding

Fog and mist occur at all times of the year and can be exacerbated by air pollution. It is useful then to have a look at air pollution in Indiana and see how our perceptions relate to the observed events. To fully appreciate the nature of air quality and pollution, the national standards are examined.

Air Quality Standards

The Clean Air Act requires that the Environmental Protection Agency (EPA) sets National Ambient Air Quality Standards (NAAQS) for six common air pollutants, labeled criteria pollutants. Exposure to elevated concentrations of these pollutants has been demonstrated to cause premature death and excess illness. The following discussion offers a description of the criteria pollutants and their concentrations in Indiana. One of the criteria pollutants, lead, is excluded from the discussion because ever since the removal of lead from gasoline it has negligible atmospheric sources, and no monitoring site in Indiana exceeded federally mandated levels during the last decade.

OF INTEREST

Five Criteria Pollutants

S. C. Pryor

Carbon monoxide:

Carbon Monoxide (CO) is an odorless and colorless gas, emitted during combustion of all carbon-based materials (for example, fossil fuels). In the United States approximately 56 percent of all carbon monoxide emissions derive from on-road vehicles, 22 percent derive from non-on-road vehicles (for example, aircraft, boats, and agricultural machinery), and the remainder derive principally from industrial processes, fossil fuel combustion in power plants, and wildfires. For regulatory purposes carbon monoxide, like all the criteria pollutants that are gases, is measured in parts per million (ppm, that is, the number of molecules of a specific gas per million gas molecules). Exposure to carbon monoxide negatively impacts human health because it bonds with hemoglobin in the blood far more readily than does oxygen, thereby reducing the oxygen supply to the body's organs and tissues. Exposure to carbon monoxide concentrations greater than 750 ppm causes lethargy in humans and eventually death.

Nitrogen dioxide:

Nitrogen dioxide (NO_2) is one of a group of nitrogen-containing gases referred to as oxides of nitrogen (NO_x). These gases are emitted from high temperature combustion of fossil fuels but are also released naturally into the atmosphere from soils and lightning events. In the United States approximately 55 percent of all anthropogenic (human-caused) emissions of NO_x come from on-road vehicles, about 22 percent derive from combustion of fossil fuels in power stations, and a further 22 percent come from industrial, commercial, or residential sources. Nitrogen dioxide is a key ingredient in reactions that produce ground-level ozone (O_3). Exposure to nitrogen dioxide concentrations exceeding those of EPA standards can cause respiratory failure, and elevated concentrations are also responsible for decreases in visibility, increase in algae formation in freshwater bodies, increased atmospheric particle formation, and acid rain effects.

Ozone:

Ozone (O_3) is a gas molecule that is beneficial in the upper layers of the atmosphere, where it is present in the so-called ozone layer that protects the earth from the sun's very short wavelength ultraviolet radiation. However, when it is produced at ground level it can cause respiratory problems in humans (particularly those who suffer from asthma) and can also stifle the photosynthesis process in vegetation causing decreases in agricultural yields. This has led some to say that ozone is good up high, but bad nearby. Unlike the other criteria pollutants, there are no direct emissions of ozone to the atmosphere; rather, it is formed in the lower atmosphere (that is, near ground level) by reactions between volatile organic compounds (VOCs, a class of carbon-containing chemicals that will readily evaporate into the atmosphere) and oxides of nitrogen.

Particulate matter:

Particulate matter (PM), or particles, is defined as particles of solid or liquid form suspended in air. Atmospheric particulate matter derives from a number of sources. It can be directly emitted into the atmosphere (for example, dust and soil may be swept into the atmosphere in part due to agricultural activities), soot may be emitted as a result of combustion of fossil fuels, and sea spray may enter the atmosphere as a result of breaking waves on the oceans. Particles may also form in the atmosphere due to chemical reactions involving VOCs derived from plant and animal activities, or inorganic gases such as sulfur dioxide produced during fossil fuel combustion. Particulate matter is a major contributor in the formation of haze and low-visibility conditions, and, when present in high concentrations, can be detrimental to human health by causing harm to respiratory and cardiovascular functions.

The EPA classifies atmospheric particles into two classes based on their diameter:

Particulate matter 2.5 ($PM_{2.5}$) and

Particulate matter 10 (PM_{10})

where the values 2.5 and 10 indicate the diameter of the particles measured in micrometers, or microns (μm), which are one-millionth of a meter. As comparison, a human hair is approximately 70 microns in diameter. For regulatory purposes particulate matter is measured as a concentration in micrograms (or one-millionth of a gram) per cubic meter of air (abbreviated to μg/m³).

Sulfur dioxide:

Sulfur dioxide (SO_2) is a toxic gas that is emitted into the atmosphere when fuels that contain sulfur (such as coal) are burned. An excess of sulfur dioxide in the atmosphere causes respiratory problems in humans and acidification of precipitation (more commonly known as acid rain) which can contribute to decreased agricultural yields and to the corrosion of metals and structures. National sulfur dioxide emissions derive mainly from fuel combustion at electrical utilities (67 percent), with fuel combustion at other industrial facilities, during metal processing, and in vehicles accounting for the majority of the remaining emissions. In the United States, sulfur dioxide emissions have decreased by about 33 percent since 1980, and no counties in the United States failed the NAAQS for sulfur dioxide during 2005. However, in some locations ambient concentrations still exceed acceptable levels as defined in United States–Canada agreements, and in 2006 some failures of the NAAQS were observed.

The Air Quality Index

To provide an accessible metric of daily air quality to the public, the EPA developed the Air Quality Index (AQI). In all large cities (that is, cities with a population above 350,000), state and local agencies are required to report the AQI to the public each day. The index has values of between 0 and 500 and, as shown in Table 4.1, higher values of the AQI indicate worse air quality. An AQI value of 100 generally corresponds to the NAAQS for a given pollutant, but to make it easier to understand, the AQI is divided into six categories. An AQI for each criteria pollutant is calculated using measured concentrations in each region, and then the highest AQI for any of the pollutants is used to report the AQI for that day. Forecasts of air pollutant concentrations are also used by the EPA to predict the AQI for the next day. These observed and forecast AQI values for each location across the United States are published each day in newspapers and are available on the internet at http://airnow.gov/.

Table 4.1. The Air Quality Index (AQI) as Used by the EPA to Disseminate Air Quality Information

Air Quality Index (AQI) Value	Level of Health Concern	Description
0 to 50	Good	Air pollution poses little or no risk
51 to 100	Moderate	Air quality is acceptable but sensitive people (e.g., asthmatics) may experience some negative health impacts
101 to 150	Unhealthy for Sensitive Groups	People sensitive to air pollution (e.g., asthmatics) may experience health effects, but the general population is unlikely to be affected.
151 to 200	Unhealthy	Everyone may begin to experience health effects
201 to 300	Very Unhealthy	A health alert may be issued. The public at large may experience serious health impacts.
301 to 500	Hazardous	Health warnings of emergency conditions will be issued. The entire population is more likely to be affected

Source: Table courtesy of S. C. Pryor

Emissions of Air Pollutants in the Midwestern USA

Emissions of air pollutants play a critical role in determining air quality. In 2005 about 141 million tons of pollutants were emitted into the atmosphere above the United States, down from 160 million tons in 2000. Since emissions of many of the criteria pollutants are dependent on human activity, it follows that the emissions patterns are spatially variable and tend to center around areas with intense human activity. The spatial patterns of air pollutants also give an indication of the relative importance of different emission sources. For example, focusing on the role of automobiles in carbon monoxide emission, one sees that the highest emission densities of carbon monoxide in the Midwest are found in counties located in or around the centers with highest population density and automobile use such as Chicago and Indianapolis (figure 4.6). On the other hand, particulate matter is emitted from a great many activities and hence emissions of $PM_{2.5}$ are more evenly distributed across the Midwest.

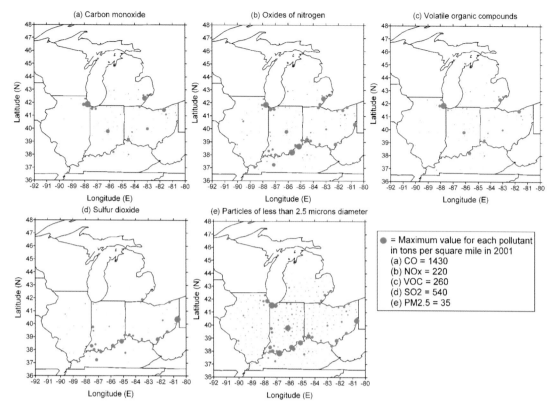

FIGURE 4.6. Emissions of five criteria pollutants in the Midwest during 2001 shown by the county in which the emissions occurred. Each frame shows a different pollutant in tons per square mile. The size of the dot changes with emission magnitude (a larger dot means a higher emission). The highest county-level emission of each pollutant is shown in the legend. Figure courtesy of by S. C. Pryor.

Monitoring of Air Quality

It follows from the emission patterns that highest concentrations of the criteria pollutants are found over and near urban areas. For this reason the monitoring of air quality in Indiana is strongly focused on the major metropolitan and industrial centers. Below is a synthesis of conditions in Indiana relative to the NAAQS based on observations from the Indiana Department of Environmental Management.

How Clean Is Our Air?

In 2007 the status of air quality in Indiana relative to the NAAQS is as follows:

- Carbon monoxide (CO): no counties where carbon monoxide measurements arecurrently conducted exhibit concentrations above the NAAQS. Two counties are designated maintenance areas because in prior years they had been designated non-attainment areas (that is, air pollutant concentrations exceeded the NAAQS). Those two counties are Lake County (northwest Indiana) and Marion County (Indianapolis).

- Nitrogen dioxide (NO_2): no Indiana counties are non-attainment areas for the nitrogen dioxide NAAQS.
- Ozone (O_3): two Indiana counties were previously designated as non-attainment areas for the 1-hour ozone standard, Lake County and Porter County (northwest Indiana), while these two and fifteen other counties across Indiana are non-attainment areas for the 8-hour ozone standard promulgated by the EPA on July 18, 1997.
- Particulate matter (PM): two Indiana counties were previously non-attainment areas for PM_{10} and hence are currently designated maintenance areas. They are Lake County (northwest Indiana) and Vermillion County (west central Indiana). Seventeen counties are non-attainment areas for the $PM_{2.5}$ standard.
- Sulfur dioxide (SO_2): no Indiana counties currently fail the NAAQS for sulfur dioxide, but five were previously non-attainment areas for this pollutant and hence are currently designated maintenance areas. They are La Porte County and Lake County (northwest Indiana), Marion County (Indianapolis), Vigo County (in the west of the state), and Wayne County (on the border with Ohio).

The EPA has a mandate to monitor, improve, and forecast air quality across the United States, and to warn and protect people from pollution levels known to be harmful to human health. Air quality is accordingly measured at multiple locations throughout Indiana. Those measurements generally indicate that the introduction of the Clean Air Act and other air pollutant control measures have led to improved air quality across the state. Nevertheless, people living in and near to the major urban areas of the state experience air pollutant concentrations that are known to be harmful to human health. The major air pollutants in Indiana are ozone and particulate matter. The NAAQS for these pollutants are exceeded in over half of the counties in which they are measured.

5

WINTER

When winter arrives, the sun is low in the sky and frigid arctic air masses, cooled during the long polar winter nights, are guided south of Indiana by an active jet stream (figure 5.1). If the arctic air masses are in place over Indiana, the sky is a beautiful blue and cold days ensue. At times, the cold air clashes with warmer and moister air from the south, and snowstorms can result. The hallmark of winter is water in its frozen state, be it snow in drifts that snarl communications or ice on trees and wires that interrupts supplies of electricity. It is a beautiful season, but one that has dangerous potential.

We can get a general guide to winter temperatures by looking at the average temperature map for January (figure 5.2). Such maps are useful in gaining some insight into winter weather in the Hoosier State, but, as is well know, there are many variations in winter days. Some are warmer and some are colder than others. It is the cold weather that often causes the greatest hardship.

The Continental Location

In chapter 3 it was noted that Indiana experiences quite hot weather during the summer. In this chapter the discussion centers on cold weather. While it is the air masses and upper air winds that determine the conditions that exist, the location of Indiana, removed from any nearby ocean, experiences a climatic control called continentality. Land masses heat much more rapidly than oceans in summer. Since this heat concentrates near the earth's surface, it rapidly cools as winter approaches. In contrast, oceans warm up more slowly but the heat is mixed to an appreciable depth so they retain heat for a much longer time. Accordingly, places close to an ocean have relatively warm winters and cooler summers. The data below show this attribute by comparing an Indiana location with one at a similar latitude but near an ocean. Looking at the data, one can see that Eureka, California, which is at 40°N, compared to Indianapolis, at 39°N, has an average January temperature almost 20°F warmer than Indianapolis, while its average July temperature is around 17°F cooler.

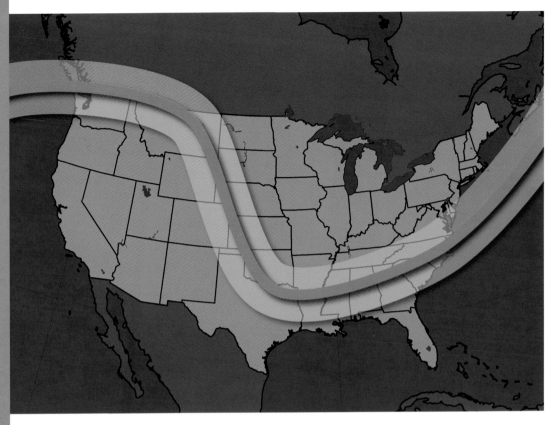

FIGURE 5.1. Average position of the winter (December–February) jet stream based upon fifty years of data from the National Oceanic and Atmospheric Administration's Earth System Research Laboratory. Figure courtesy of Cameron Craig.

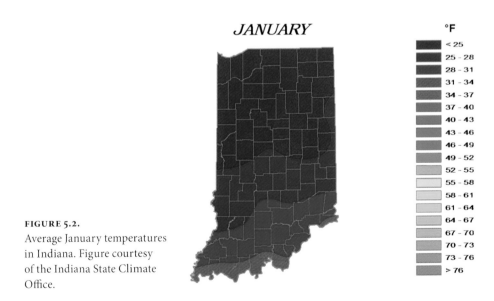

JANUARY

°F

- < 25
- 25 – 28
- 28 – 31
- 31 – 34
- 34 – 37
- 37 – 40
- 40 – 43
- 43 – 46
- 46 – 49
- 49 – 52
- 52 – 55
- 55 – 58
- 58 – 61
- 61 – 64
- 64 – 67
- 67 – 70
- 70 – 73
- 73 – 76
- > 76

FIGURE 5.2.
Average January temperatures in Indiana. Figure courtesy of the Indiana State Climate Office.

City	Latitude	January Minimum	January Maximum	January Average	July Minimum	July Maximum	July Average
Eureka, Calif.	40°N	41	55	48	53	63	58
Indianapolis, Ind.	39°N	21	37	29	64	86	75

Not only is Indiana located away from an ocean, it is also located on a continent that has its uplands and mountain systems oriented in a general north-south direction. There is no mountain barrier between the Arctic Ocean and the Gulf of Mexico. The entire area is open to frigid air masses from the north and balmy breezes from the south. The groundwork is set for contrasting temperatures, and giant clashes between air masses of different properties.

Winter Precipitation

The average amount of snowfall in Indiana varies appreciably from north to south (figure 5.3). In an area south of the Great Lakes, the annual snowfall averages fifty inches; in the south some areas receive less than ten inches annually. The average date of the first measurable snowfall in Indiana is November 19, and the average date of the last is March 30.

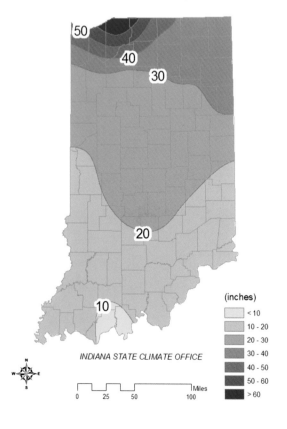

MEAN WINTER SNOW
(1974 - 2003)

(inches)
< 10
10 - 20
20 - 30
30 - 40
40 - 50
50 - 60
> 60

INDIANA STATE CLIMATE OFFICE

Miles
0   25   50   100

FIGURE 5.3.
Average snowfall amounts in Indiana, 1974–2003. Figure courtesy of the Indiana State Climate Office.

Snowmaking in Indiana

When examining the snowfall records (Table 5.1), it is seen that they are dominated by stations in the northern tier of Indiana. Note that most of the records are related to a single year at La Porte. As indicated below, there have been questions raised about these records, although 1958 was a record-making snowfall year.

Many people are interested in experiencing a Christmas with snowy scenes. In Indiana, the chances of a "white Christmas" again reflect the north-south snowfall distribution. In the areas just south of Lake Michigan, there is a 50 percent chance that Christmas

Table 5.1. Indiana Snowfall and Snow Depth Extremes

	Snow amount (inches)	Location COOP station number	Location Station name	Ending date	Number of years of non-missing data	Data period analyzed
Greatest daily snowfall	22.0	121256	Cannelton	12/23/2004	22	1972–2006
Greatest 2-day snowfall (snowed both days)	29.0	128187	South Bend	01/31/1909	89	1893–2006
Greatest 3-day snowfall (snowed all 3 days)	32.0	128187	South Bend	01/31/1909	89	1893–2006
Greatest 4-day snowfall (snowed all 4 days)	33.0	124837	La Porte	02/18/1958	58	1897–2006
Greatest 5-day snowfall (snowed all 5 days)	36.0	124837	La Porte	02/19/1958	58	1897–2006
Greatest 6-day snowfall (snowed all 6 days)	37.0	124837	La Porte	02/19/1958	58	1897–2006
Greatest 7-day snowfall (snowed all 7 days)	37.5	124837	La Porte	12/21/1951	57	1897–2006
Greatest monthly snowfall total	59.5	124837	La Porte	02/1958	55	1897–2006
Greatest August–July snowfall total	125.6	124837	La Porte	1978	28	1897–2006
Greatest daily snow depth	38.0	124837	La Porte	02/18/1958	45	1899–2006

Table 5.2. Some Winter Definitions

Winter Event	Definition
Blizzard	Winds of 35 mph or more with snow and blowing snow reducing visibility to less than ¼ mile for at least three hours.
Blowing Snow	Wind-driven snow that reduces visibility. Blowing snow may be falling snow and/or snow on the ground picked up by the wind.
Snow Squalls	Brief, intense snow showers accompanied by strong, gusty winds. Accumulation may be significant.
Snow Showers	Snow falling at varying intensities for brief periods of time. Some accumulation is possible.
Snow Flurries	Light snow falling for short durations with little or no accumulation.

will be white, while in southern Indiana the probability is less than 10 percent, less than one year in ten.

Like all major seasonal events, winter has its own vocabulary. Table 5.2 provides a key to some of the main terms. At the top of the list is the blizzard, a storm that often has serious consequences for stranded travelers. However, snow is not the only form of winter precipitation and in some instances it is not always the most dangerous.

The type of winter precipitation that occurs depends upon the temperature structure of the lower atmosphere (figure 5.4). If the temperature from the earth's surface upward is below freezing, then snow can occur. The formation of sleet—ice pellets that bounce when striking the ground or produce a tapping sound when hitting a window—occurs when a falling snowflake melts as it passes through a layer of warmer air. When it passes through a thick layer of air that is below freezing, the melted drop freezes in the form of an ice pellet (rather than a snowflake) and hits the ground as sleet. A very dangerous condition occurs when a shallow layer of below-freezing air is close to the ground. In this case, the water droplets do not freeze in the air, but rather freeze on immediate contact with the surface. A veneer of ice is formed that turns roads into skating rinks and accumulates on trees to break limbs, which may block roads and bring down power lines.

The Lake Effect

Craig Clark

Seasonal snowfall in much of northern Indiana is increased due to lake-effect snow. This is the result of cold air passing over warmer water from which it derives moisture, and then depositing the moisture as snow on the lee shore of the lake, the shoreline that has the wind blowing toward it (figure 5.5a). The primary lake-effect region occurs from about La Porte to South Bend, with roughly

It is often noticed that after a snowfall, it seems much quieter than usual. The deeper the snow the quieter it appears. This phenomenon is a function of the way that sound is propagated. Just like an acoustic tile, snow absorbs rather than reflects a large percentage of the sound waves that strike its surface. As the snow becomes older and more compact, its ability to absorb sound decreases as does the so-called sound of silence.

Just as the age of snow influences its sound properties, so too does its temperature. If you are walking on snow-covered pavement you may notice that with each step, a squeaking sound occurs. For this to happen the snow temperature must be below 14°F, a temperature at which the weight of the boot will not melt the ice crystals, but rather crush them; it is the crushing of the ice crystals that causes the typical squeaking sound. At temperatures above 14°F the pressure of the boot partially melts the snow, and no sound results.

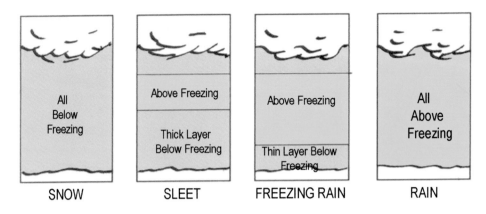

FIGURE 5.4. The structure of the atmosphere below the clouds determines the type of precipitation that occurs in winter.

FIGURE 5.5A. Schematic diagram showing how lake-effect snow occurs in the lee of the lake. Adapted from Morgan and Moran (1997).

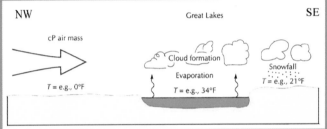

FIGURE 5.5B. Mean seasonal snowfall (inches) in the lees of the Great Lakes. The heaviest Indiana snowfalls are about sixty inches in the snowiest areas of the north. Adapted from Morgan and Moran (1997).

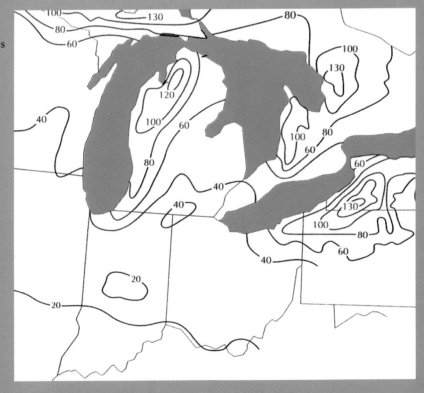

half of the average seasonal snowfall occurring due to lake-effect snowstorms (figure 5.5b). Although locations in Lake and Porter Counties occasionally have heavy lake snows, they are less common.

As seen, lake-effect snowstorms occur during the winter as a result of cold air flowing over the relatively warm water of Lake Michigan. Although the lake is warmer earlier in the season, the frequency of Indiana lake-effect snow peaks in December and January. While snowfall totals can be very impressive, the most frequent events have less than six inches of reported snowfall.

Lake-effect events most commonly occur following the passage of a low pressure system, often an Alberta Clipper, from the northwest. It was noted earlier that as the cold air flows over the warm lake, the air is heated by turbulent mixing from below, generating a very unstable layer of air at low levels. Evaporation adds water vapor to the air mass, which moistens the air mass and lowers the cloud base. Snow bands form over the water and move inland, where the additional roughness of the surface enhances snowfall intensity.

How cold does the air have to be? If the difference between the lake temperature and the temperature of the air at about 4800 feet (1300m, or where air pressure is 850 millibars [mbs]) above the surface is large enough, significant snowfall may occur. A difference between two temperatures is expressed in scientific shorthand as the Greek letter delta (Δ) and given as ΔT. The commonly cited threshold is a ΔT of 13°C. However, nearly all of the significant—that is, accumulation of at least six inches—lake-effect events in Indiana occur with ΔT at least in the upper teens (figure 5.6). While greater instability is clearly favorable for a heavier lake-effect snowfall, many cases with ΔT exceeding 18°C have only modest snowfall.

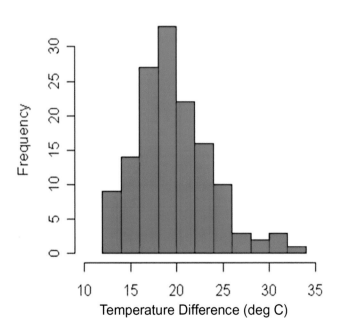

FIGURE 5.6. Temperature difference (Delta T) in °C between the lake surface and the 850 mbs pressure level estimated for each archived Indiana lee of the lake event. The graph shows that the most frequent snow events occur when the temperature difference is approximately 20°C. Figure courtesy of Craig Clark.

A long fetch, or distance the air crosses over the lake, is critical for impressive lake-effect snowfall. A shorter fetch will minimize the air mass modification. Most events occur with winds from the northwest, which carry the snow toward eastern La Porte and St. Joseph Counties. These events usually have multiple wind-parallel bands, which evolve quickly over time. Forecasting the precise location of the heaviest snow is quite challenging, in large part because the wind direction and speed change with height and may also change over time during the event.

With due northerly winds, the very long fetch across Lake Michigan lets the air mass accumulate moisture for a long time and thus favors heavy lake-effect snow in northwest Indiana (typically Porter County or La Porte County). These comparatively rare events are usually defined by a single, intense mid-lake band. These mid-lake events are also strengthened by the development of land breeze circulation. With a land breeze, the low-level flow from the land to the lake converges at the mid-lake snow band, with enhanced rising motion and hence more cloud development.

In some rare cases, very weak large-scale flow and land breeze development lead to the development of a vortex over the lake. These events do not generally produce much snowfall, but they are visually impressive on radar and satellite images.

What are the inhibiting factors that reduce amounts of snowfall? Wind direction is likely the biggest cause of snowfall forecast error, since a small error in wind direction can make a large change in projected snowfall location and amount. One key inhibiting factor is the inversion height (that is, the height above the surface at which temperature starts to increase with height). If the inversion is too low (approximately 1500 m is a typical guideline), snowfall may be inhibited. Assessment of this forecasting guideline is complicated, since the lake heats the air and can significantly raise the height of the inversion as the air passes over the lake. In rare cases, very high wind speeds can also inhibit lake-effect snow totals by reducing the amount of time that the air mass spends accumulating moisture over the lake. Very dry upstream air masses can also inhibit snowfall totals in some cases; this does not appear to be a strong factor, perhaps because a lake's long fetch may overcome this inhibition.

Cold Weather

Not every winter day is freezing cold. But when the temperature is well below freezing for days on end, it is usually because a cold wave, or outburst of polar air, has swept into town. A cold surge begins over the vast plains of Canada stretching into the Arctic Circle. During the long winter nights the snow-covered tundra cools in the clear arctic air and a great dome of cold air forms. Surface temperature falls as low as 40 degrees below 0°F. The dome of cold, heavy air spreads out, pushing against the warmer air to the south. Between the two air masses a front forms. This cold front is the forward edge of the crisp and crystal clear arctic air mass. Although heating by the warmer ground raises its temperature, the arctic air is still very cold: the temperature of the air may be 20°F on the cold side of the front and 50°F on the warm side.

As the front passes the wind swings around, now blowing from the northwest. In a few minutes the temperature may drop abruptly from 50 to 30°F. As more cold air is brought in by the northwest winds, the temperature continues to fall. The front has moved beyond Indiana, but behind it cold air still pours down from Canada. When night falls, the ground radiates heat through the dry, clean arctic air and the temperature of the air falls rapidly. By

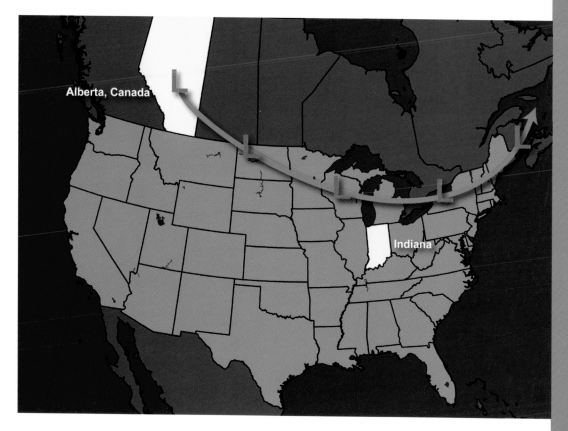

FIGURE 5.7. Path of a typical Alberta Clipper. Figure courtesy of Cameron Craig.

morning the frigid temperature causes the water vapor in our breath to condense, making miniature clouds near our faces.

A similar deep freeze may be produced by an Alberta Clipper, so named because it originates in the province of Alberta in Canada (figure 5.7). The term "clipper" is used to provide an image of a very rapidly moving body of air, leading to an often drastic, sudden temperature decline. It will be recalled that this low pressure system is important in lake-effect snowfalls.

As illustrated in the examples given below and data provided in chapter 8, winter in Indiana can be cold! But the cold of winter is tempered by the warmth of home. The images of pioneers huddled around an open wood-burning fireplace are long gone. Today central heating is used by most people to heat their homes, and this heating requires a fuel source provided by oil, natural gas, and electricity companies. The amount of fuel used is estimated using the concept of heating degree days (HDDs).

Heating Degree Days

Heating engineers found that when the mean daily outdoor temperature is below 65°F, most buildings require interior heating for comfort and this became the basis for working out how much heating may be required. HDDs are accumulated by subtracting the average daily temperature from 65°F. Here is an example of how the HDDs are calculated:

With a daily high temperature of 70°F and a low of 50°F, the average is 60°F

The average (60°F) is subtracted from 65°F to give 5 HDDs

Suppose, then, that the next day has a mean daily temperature of 58°F, then an additional 7 (65–58) HDDs will be accumulated

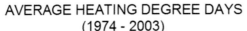

AVERAGE HEATING DEGREE DAYS
(1974 - 2003)

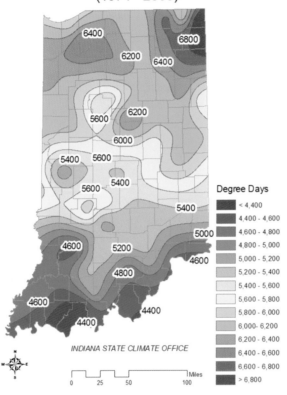

FIGURE 5.8.
Average heating degree days in Indiana, 1974–2003. Figure courtesy of the Indiana State Climate Office.

By keeping a running total of HDDs throughout the heating season, fuel distributors and power companies can anticipate the general amount of fuel to be delivered or power to be generated. Large annual differences across the United States are seen in these accumulated averages. In central North Dakota, for example, some 9000 HDDs occur, while in central Florida 500 HDDs accumulate. Clearly, heating costs will be much greater in the colder northern states. As can be seen in figure 5.8, HDDs in Indiana range from 4400 in the south to more than 6800 in the northeast. Heating costs in winter for the northeast will be accordingly higher. Fortunately, the cost of cooling in summer is exactly opposite, with higher costs in the southern parts of the state (see figure 3.11).

Windchill

In the discussion of summer heat, it was noted that if more heat is gained by the body than is lost, then body temperature rises; when this happens, the body's internal mechanisms are triggered to bring body temperature back down into its typical range. Conversely, if more heat is lost by the body than is gained, then body temperature falls; here, too,

Wind Chill Chart

Wind (mph) \ Temperature (°F)	40	35	30	25	20	15	10	5	0	-5	-10	-15	-20	-25	-30	-35	-40	-45
5	36	31	25	19	13	7	1	-5	-11	-16	-22	-28	-34	-40	-46	-52	-57	-63
10	34	27	21	15	9	3	-4	-10	-16	-22	-28	-35	-41	-47	-53	-59	-66	-72
15	32	25	19	13	6	0	-7	-13	-19	-26	-32	-39	-45	-51	-58	-64	-71	-77
20	30	24	17	11	4	-2	-9	-15	-22	-29	-35	-42	-48	-55	-61	-68	-74	-81
25	29	23	16	9	3	-4	-11	-17	-24	-31	-37	-44	-51	-58	-64	-71	-78	-84
30	28	22	15	8	1	-5	-12	-19	-26	-33	-39	-46	-53	-60	-67	-73	-80	-87
35	28	21	14	7	0	-7	-14	-21	-27	-34	-41	-48	-55	-62	-69	-76	-82	-89
40	27	20	13	6	-1	-8	-15	-22	-29	-36	-43	-50	-57	-64	-71	-78	-84	-91
45	26	19	12	5	-2	-9	-16	-23	-30	-37	-44	-51	-58	-65	-72	-79	-86	-93
50	26	19	12	4	-3	-10	-17	-24	-31	-38	-45	-52	-60	-67	-74	-81	-88	-95
55	25	18	11	4	-3	-11	-18	-25	-32	-39	-46	-54	-61	-68	-75	-82	-89	-97
60	25	17	10	3	-4	-11	-19	-26	-33	-40	-48	-55	-62	-69	-76	-84	-91	-98

Frostbite Times: 30 minutes / 10 minutes / 5 minutes

$$\text{Wind Chill (°F)} = 35.74 + 0.6215T - 35.75(V^{0.16}) + 0.4275T(V^{0.16})$$

Where, T = Air Temperature (°F) V = Wind Speed (mph)

Effective 11/01/01

FIGURE 5.9. The Wind Chill Chart. Figure courtesy of the National Oceanic and Atmospheric Administration and the National Weather Service.

internal mechanisms bring body temperature back up into its typical range. The term "hypothermia" refers to the condition where core body temperature is reduced below a safe range, while the term "hyperthermia" refers to the condition where core body temperature is raised above a safe range. Both conditions can be life threatening.

Cold conditions cause a reduction of blood flow to the periphery—for example, hands and feet—so that less heat is lost from the body. While this reduced outward flow conserves heat to maintain the core temperature of the body, the reduced blood flow to the extremities can have dire effects. Under extreme conditions, frostbite can occur, resulting in frozen tissue and cell destruction.

Shivering is another response to cold conditions. Its function is to increase the metabolic rate. While this certainly occurs, shivering is not a highly efficient process. It brings more blood to the surface layers of the body but this increases potential heat loss by radiation and convection to the surrounding environment. Uncontrolled shivering eventually causes difficulty in breathing, muscular rigidity, unconsciousness, and, unless treated, death.

Just as an index has been derived for heat stress—the Heat Index—so too has one been derived for the influence of low temperatures on the human body. The term "windchill" was coined by Antarctic explorer Paul A. Siple to describe the cooling power of wind for various combinations of temperatures and wind speeds. Later, he provided a method to calculate windchill. This index, developed from experiments conducted in Antarctica, measures the rate at which water freezes at various temperatures and wind speeds.

The original application of the Siple formula was introduced to predict conditions that cause frostbite. A modified windchill formula, which included the effect of clothing, was later introduced. Because a clothed person was considered, the variables of breathing and heat transfer through clothing were incorporated. The model assumes a healthy adult of height 5ft 6in (1.7 m) walking outdoors at 3 mph (1.33 m/sec). To maintain thermal equilibrium, the amount of heat loss must not exceed the heat generated. The balance is achieved by wearing an appropriate thickness of clothing. The present windchill advisories are based on the sensation of cold felt by the majority of people. Figure 5.9 outlines how windchill is derived.

Some Recent Severe Winters

The Midwestern Regional Climate Center (see chapter 7) provides excellent accounts of all types of weather events across our region. The following summaries of recent winter events are drawn from their descriptions.

January–February, 1996

One of the coldest weather events of the twentieth century occurred on these dates in the Midwest. Daily low temperature records were set especially in the northern parts of the area (figure 5.11). The following data show this, and while Indiana temperatures did not attain the record proportions of Illinois and Wisconsin they certainly ranked high on the list of twentieth-century cold spells.

Date of Cold Outbreak	Indiana Average Temperature (°F)
Jan. 18 to Jan. 21, 1994	-5.4
Dec. 21 to Dec. 24, 1989	-4.7
Dec. 23 to Dec. 26, 1983	-3.8
Jan. 16 to Jan. 19, 1977	-3.0
Jan. 19 to Jan. 22, 1984	-1.0
Feb. 02 to Feb. 05, 1996	-0.1

The coldest average temperature recorded occurred in February 1899 with an average of 6.1°F; the second coldest occurred as recently as December 2000.

December 2000

The Midwestern Regional Climate Center reports that our region experienced its second coldest December in the 106 years that observations have been recorded. The December 2000 average temperature was 14.3°F, only slightly warmer than the 1983 record of 13.9°F. South Bend, together with stations in Illinois and Kentucky, broke their all-time cold records for December.

The center of unusually cold conditions was in Iowa, Missouri, and Illinois, where temperatures averaged 12–15°F below normal. Even in relatively warmer locations such as northern Michigan, temperatures were still more than 5°F below normal. During December 2000, a strong high pressure ridge dominated the western coast of North America, and this enhanced the passage of the north-to-south delivery of very cold air. December 2000 was also characterized by heavy snowfall due in part to these sustained low temperatures (figure 5.10).

December 22–23, 2004

At this time the worst winter storm on record in southern Illinois occurred. The storm then moved eastward influencing Indiana, Kentucky, and Ohio. Snowfall amounts from

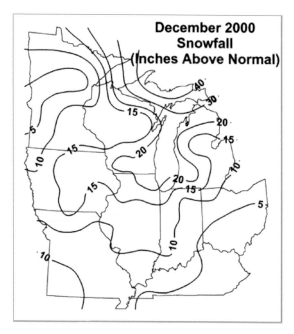

**December 2000
Snowfall
(Inches Above Normal)**

FIGURE 5.10.
Snowfall amounts above normal
in the Midwest in December 2000.
Figure courtesy of the Midwestern
Regional Climate Center.

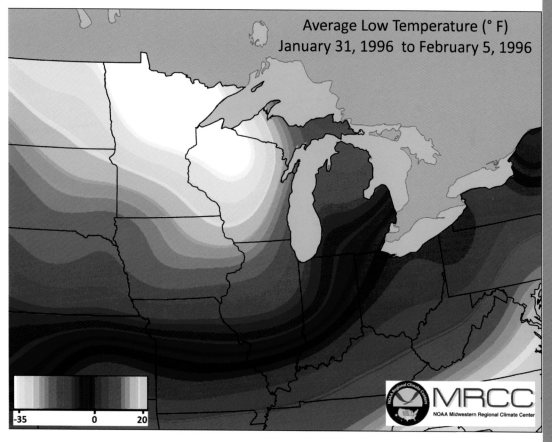

Average Low Temperature (° F)
January 31, 1996 to February 5, 1996

FIGURE 5.11.
Average low temperature (°F), January 31, 1996, to February 5, 1996. Figure courtesy of the
Midwestern Regional Climate Center.

the storm that lasted thirty hours at many locations reached twenty-nine inches, and more than six inches fell over a 137,600-square-mile area. Snowfall totals set new records across southern Illinois, the southern half of Indiana, and western Ohio. This prolonged, enormous storm system also produced a major ice storm along its southern edge in Kentucky and Ohio. This storm was extremely destructive and caused nine hundred million dollars in losses and costs. A report about this storm from the Midwestern Regional Climate Center tells us:

> All aspects of transportation were affected, and the impacts were extreme because the storm occurred at a time of extensive pre-holiday travel. Traffic was paralyzed on numerous interstate highways, and thousands of persons were stranded for 6–36 hours in the bitter cold. Hundreds of airline flights were cancelled or delayed, and trains were halted at several locations. Thousands of vehicular accidents led to numerous injuries, and 17 persons died as a result of the storm.

6

MONITORING THE WEATHER

In these days of rapid communication and instant information it is often forgotten that in order to provide accurate information numerical data are needed. This is particularly true as far as weather is concerned. Consider an average morning when many people are getting ready for work or school. One piece of information they look for is the day's weather forecast. For some it may just help them to decide what to wear. For others, be they farmers or outdoor construction workers, the information may decide their plans for the day.

But long before any weather map or forecast is available, a great deal of work is required of meteorologists. Data from surface stations, weather balloons, buoys at sea, and satellites in place over the earth are used to construct the current conditions—and these are used to produce a forecast. The forecast may then be available through television weather programs, a very important component in public awareness of atmospheric conditions. These issues are considered in this chapter.

But before embarking on a discussion of modern weather forecasting, it is both instructive and interesting to consider the older, and perhaps less reliable, method of forecasting—weather sayings.

Weather Sayings

Ronald L. Baker

"If you don't like Indiana weather, just wait five minutes, and it'll change"—or so goes a Hoosier folk saying about the weather. Of course, the same saying is found throughout the United States about other states and regions. In fact, most Hoosier weather sayings are common not only throughout the United States but also throughout most European countries and other countries where Europeans have settled. But that is what makes these sayings folklore; if there are different versions of a verbal expression, then you can be almost certain that it is folklore. Moreover, the various traditional genres dealing with weather—mainly legends, personal experience narratives, tall tales, sayings, proverbs, and superstitions (also called "folk beliefs")—also are ubiquitous.

When most people think of weather lore, though, they do not think of legends, personal experience stories, and tall tales; rather, they think of sayings, which abound in Hoosier folk culture. Some weather sayings are proverbs; some are ordinary sayings; and many more are superstitions. In weather lore it is hard to distinguish proverbs from superstitions, however, because weather proverbs usually mean just what they say and they lack metaphorical shifts in meaning like other proverbs. As a matter of fact, it can be argued that most weather proverbs are really superstitions since they are interpreted literally, not metaphorically. However, formularized weather sayings with definite form and content that employ rhyme and meter, like "April showers bring May flowers," or other poetic qualities may be classified as proverbs. Simple prosaic expressions about the weather that are not signs and do not really forecast, like the one that opens this section, are ordinary sayings. Folk expressions, proverbial or prosaic, that are signs or predict the weather, such as "A hog with a stick in its mouth is a sign of rain," are superstitions. Regardless of how they are classified, there are a great many weather sayings and superstitions in Indiana because weather is of vital interest to Hoosiers, especially to travelers, farmers (who once constituted a large percent of the population), and other outdoor workers and sports people who on a daily basis are exposed to all kinds of weather. Travelers are especially interested in the weather, and a popular rhyme for predicting rain in Indiana goes:

> Evening red, morning gray,
> Sets the traveler on his way.
> Evening gray, morning red,
> Brings the rain down on your head

But nearly as many Hoosiers hold the opposite view:

> Morning red, evening gray;
> Sets the traveler on his way.
> Morning gray, evening red;
> Brings rain upon his head.

Above and Below the Sky

Hoosiers not only have used the sky to predict the weather, as in the above rhyme, they also have relied on the sun, moon, stars, humans, animals, birds, insects, and plants to predict the weather. Older folks may be better indicators of weather than young folks, though, since aching corns and bones often indicate rain, as in this Hoosier belief: "When an old person's rheumatism bothers them, it means it is going to rain." Since weather lore results from long experience and close observation of nature by those who work, play, and travel outdoors, some superstitions accurately forecast the weather, though short-range predictions generally are more accurate that long-range predictions. Taken as a whole, however, most weather superstitions are inaccurate and are of more value in folkloristic than in meteorological study.

Among other things, weather superstitions deal with the seasons, rain, drought, wind, tornadoes, lightning, thunder, hail, snow, frost, and ice. Most superstitions about the seasons have to do with spring and winter, although there are a few about summer and autumn. Animals and birds are often indicators of weather. A Hoosier says, "On a specific day in February, which I can't remember right now, the groundhog comes out of his hole to predict the weather. If it is a gloomy day and the groundhog is not able to see his shadow, he stays out because winter is over. If he sees his shadow he gets scared and runs back in the hole for six more weeks of winter." More to the point, one informant observed, "If the groundhog sees its shadow on February 2, spring is six weeks away." Some Hoosiers, as some folk in other states, are convinced that February 14, not February 2, Candlemas Day, is Groundhog Day. According to one informant, "February 14 was always Groundhog Day until that Democrat president, Roosevelt, I guess it was, got in office. He changed it to February 2 because he wanted it that way. He didn't ask the people at all." Candlemas Day, marking the midpoint of winter, however, is a good day for predicting the weather, as a Hoosier folk rhyme suggests:

If Candlemas Day be mild and gay,
Go saddle your horses and buy them hay.
But if Candlemas Day be stormy and black,
It carries the winter away on its back.

Frogs are just about as popular as the groundhog as indicators of seasons, the most popular belief stating that the frog's pond must freeze over three times before frosts are over: "Counting from the frog's first holler in the spring, he will still have to look through ice three times before summer." More generally, "When frogs croak / Winter's broke."

Birds, too, are common indicators of spring: "Spring is here / When robins appear" and "There will be no more frost in the spring / Once you hear the whippoorwill sing."

Whether Easter is early or late also determines when spring arrives, according to Hoosier superstitions: "Early Easters mean early springs, and late Easters mean late springs."

Plants and storms also indicate the arrival of spring: "There will be cold weather until the snowball bushes bloom." "A thunderstorm in February brings an early wet spring," but "Thunderstorms in March mean a late spring."

Winter has inspired more weather superstitions than any other season, and many of these deal with predicting the severity or length of winter. Animals are especially good at indicating the severity winter. One Hoosier says, "They used to say when horses played, ran with their tails up in the air, it was going to be a bad winter. I notice that comes true. Every time you see them running through the field, it does turn cold." Generally, the thickness of fur an animal has in the fall predicts the severity of the coming winter. For instance, "When the cattle get thick, long hair, it will be a long, cold winter."

Thick hair on a horse, dog, wild animal, or cattle in the fall also predicts a long and/or cold winter, and the thicker the coat, the harder the winter will be. Some Hoosiers claim that "when woolly worms wear a heavy coat, there will be a cold winter," but more often they say that "If you find a woolly worm, you can see how cold the winter is going to be by the amount of black on the worm. The more black on the worm, the colder it will be." Some say that "The black part of woolly worm represents the bad part of a winter. If he has lots of black on him that means it will be a cold bad winter. If he is black in the middle with orange on each end that means it will be mild at the first and last of the winter and bad in the middle." As one Hoosier explains, "In the fall when the woolly worms cross the road; if it is light colored we will have a mild winter, if it is black we will have a very severe winter, if it is black on both ends and light in the middle we will have early cold, a mild midwinter, and a cold late winter." Also, according to Hoosier folklore, "When fish move to deep water in the fall, it means cold winter," and "If the squirrels gather their nuts early in the fall, it's going to be a cold winter." It will also be a bad winter if "ants build their hills high" and "hornets build their nests high up in the trees."

As one older Hoosier generalizes, "The behavior of animals will foretell the kind of weather approaching," and animals and birds are common indicators of all kinds of weather, not just winter. For instance, one informant says, "If a cat washes herself calmly and smoothly, the weather will be fair. If she washes herself against the grain, foul weather is in store. And if she lies with her back to the fire, there will be a squall." Cows and pigs also are good indicators of weather. "If you see cattle out in a pasture and all their butts are pointing the same way, that means it is going to rain," according to one Hoosier. Another informant claims that "Farmers believe if all the cattle in the field move or stand in the corner, bad weather is coming." If pigs have broom sage, straw, or cobs in their mouths to make a bed, it "is a sure sign that it's going to rain." Birds huddling on the ground or flocking around homes also indicate bad weather. One older Hoosier claims that "If an owl hoots on the north side of the hill there will be good weather. If he hoots on the south side, there will be bad weather."

Plants also are indicators of winter. For instance, "If the skin of an onion is thin, there will be a mild winter. If the skin is thick, there will be a bad winter." Corn also is a very popular indicator of a bad winter: "If corn shucks and silk grow thicker and tighter around the ears of corn, it will be a harsh winter." One informant explains, "If the husk on an ear of corn are loose and don't fit tight, it's a sign of fair weather, but if the husks fit tight over the ear, it's a sign of bad weather, and the husk is tight to keep rain and snow off the ear." Generally, when bark, skin, peelings, or husks of any vegetable or fruit are thick, there will be a cold winter. An especially interesting popular belief involves persimmon seeds: "Split a persimmon seed and see if there is the shape of a spoon, knife, or fork. A spoon means a lot of snow. A knife means cold, and a fork means an 'open' or snowless winter."

Since rain is vital to plant and animal life, by far the largest group of weather beliefs deals with rain. Since antiquity, quarter days, falling at the change of seasons, have been a good time for divination and prediction, and the Easter season, the modern functional equivalent of the old spring quarter festival, is a good time for predicting rain in Hoosier folklore. A very widespread belief is that "If it rains on Easter Sunday, it will rain for seven Sundays." Other beliefs clustering around Easter are: "If it rains on Palm Sunday, it will rain on the following seven Sundays." "If it rains on Good Friday, then it will rain seven weeks more." And "If it rains the Monday after Easter, it will rain for seven straight weeks."

Rain on Sunday

The days of the week also are important in forecasting rain: "If it rains on Sunday, there will only be one fair day in the week." "If it rains on the first Sunday of the month, it will rain for seven Sundays," or "If it storms on the first Sunday of the month, there will be only one pleasant Sunday during the month."

Some Hoosiers claim that if it rains on Monday, it will rain every day of that week, but others say it will rain only four more days, three more days, or two more days, with most informants going for three more days. At least one person declares that "If it rains on Thursday it will rain the rest of the week." Rain on the first day of a month also portends more rain. A couple of informants say that if it rains on the first day of a month, it will rain

FIGURE 6.1. In folklore the banding of the woolly worm is used to estimate the severity of winter. Figure courtesy of Cold Spring School, New Haven, Conn. (http://www.coldspringschool.org).

fifteen days that month, but others claim it will rain twenty days, twenty-one days, seven weeks, or that it will just be a rainy month. One Hoosier says that "The weather on the first day of the month rules the first half of the month, and the weather on the second day rules the second half of the month." Others maintain that June 1 and July 15 are good days for forecasting rain: "If it rains on the first day of June, it'll rain for 21 days in June." And: "A shower on July 15th means forty days of rain." The latter is derived from a long-held belief that if it rains on Saint Swithin's feast day, July 15, it will rain for forty days.

There also are many Hoosier folk beliefs concerning drought, again most having their origin in European folklore. Like other weather beliefs, the moon affects drought. As one Hoosier explains, "The moon governs drought. If a long drought is going to break, it will happen in the last quarter of the moon." Other informants claim that the farther the moon is to the south, the greater the drought, while another says that if the moon remains low in the southern sky, prepare for a severe drought. The wind is also an indicator of drought. One informant contends that if the wind blows from the south on March 20, there will be a summer drought. Or, generally, according to an Indiana weather proverb: "Wind in the south / Brings a drought." According to another weather rhyme, lightning has an effect on drought: "Lightning in south / Is a sign of drought."

Most weather sayings in Indiana reflect an earlier time when people spent more time outdoors, but since folk culture is dynamic, new beliefs arise. For instance, one Hoosier informant claims that "Bad weather was caused by all those rockets going through the atmosphere, punching holes in the sky." That, in short, seems to be the stuff that weather beliefs are made of.

Today we do not use weather sayings as the main source of weather. Instead, we call upon the work of the National Weather Service in Indianapolis and the weather programs provided by television meteorologists and backed by many modern tools from

Since the launching of the first specialized weather satellite, Tiros 1, in 1960, images of earth and its cloud cover have become a common feature of television weather programs. The changes that have occurred since the launching of the early satellites concern the orbit and the sophistication of the sensors aboard the satellites. Two fundamental orbits are used, the high-altitude geostationary orbit and the low-level polar orbit.

Geostationary satellites are placed in orbit such that the satellite circles the earth at the same rate as the equatorial spin of the earth so that it stays in the same point above the earth. To maintain the same rotational period as the earth, a satellite is in a fixed position at 22,236 miles (35,786 km) above the equator. An orbit permits a full disk image of the globe to be derived every half hour or so, and because of its high altitude the satellite can view a huge area (the viewing area of a satellite is called its footprint). An advantage of the geosynchronous orbit is that it requires no tracking; that is, just like your home television satellite signal, it is not necessary to constantly adjust the receiver to the satellite position.

The GOES satellite systems, first launched in 1975, were the first operational geostationary satellites. Currently the United States has two in operation: GOES-East over the Amazon River provides most of the United State's weather information, while GOES-West is over the eastern Pacific Ocean. Most news media use geostationary photos as single images or as loops in their daily weather presentations

Doppler radar to satellite imagery. Later, it will be demonstrated how many individuals, as part of the Cooperative Weather Program, are important in collecting weather data, data that are used to construct an image of Indiana's climate over time.

The National Weather Service

John Kwiatkowski

Forecasting Indiana's weather involves input from all over the world, including national centers within the United States. However, the final forecasts for Indiana from the National Weather Service (NWS) are produced locally by a handful of people (figures 6.2 and 6.3).

Worldwide data collection makes modern weather forecasting possible. What happens over Siberia today can affect Hoosier weather next week. Global weather observations are coordinated by the World Meteorological Organization (WMO), an affiliate of the United Nations. Data gathered under WMO protocols constantly stream into several national centers, including the National Meteorological Center (NMC) of the United States. Normally, national centers handle aspects of weather forecasting done most efficiently on a large scale—for instance, computer generated predictions for large areas. The NMC is divided into subcenters with particular specialties. For example, the Storm Prediction Center concentrates on severe convective weather and issues the well-known tornado and severe thunderstorm watches. The Tropical Prediction Center handles hurricanes and other aspects of low-latitude weather. The Environmental Modeling Center concentrates on computer generated forecasts. There are other centers, but these are the ones most likely to influence Hoosiers.

The Work of Local Offices—the FOs

All the agencies mentioned above feed information into local offices, known as Forecast

Offices, or FOs. Indiana is typical for a state of its size in that it has two FOs. One is in Indianapolis, the other in North Webster. Some counties on the periphery of Indiana are served by offices in neighboring states.

Data Collection

FOs have the primary forecast responsibility for the areas they serve. They also have the less widely known but equally vital job of data collection from their County Warning Area. Most offices administer and service a network of cooperative (COOP) observers within their County Warning Area. These observers are mainly located in small towns and may number in the hundreds. They normally make daily reports of high and low temperature, as well as rainfall, at their location. The COOP-derived information is the backbone for almost all climate studies done in the United States. A modern national weather service could not exist without them. COOP program maintenance is primarily done by the FO's staff of hydrometeorological technicians (HMTs).

As the data collection experts of an FO, HMTs have many responsibilities beyond the COOP program. They determine whether the office radar is working optimally given ambient conditions, and not detecting false returns. HMTs collect and perform quality control on surface data of all kinds, including special climate information, and real-time severe weather reports. The data collection aspect of FO operations is the foundation for everything else, since nothing can be forecast without observations.

Public Services

Much of the tremendous amount of meteorological data that is constantly generated is collected in the AWIPS—Advanced Weather Interactive Processing System—computers of each FO, where it becomes the basis for forecasts. Most forecasting starts by using the output from the various national centers as a first guess. On a large scale, predictions from those centers are usually correct and

Polar orbiting satellites are placed in orbit some 497 miles to 932 miles (800 km to 1500 km) above the surface. The National Oceanic and Atmospheric Administration (NOAA)-class satellites have orbits that pass very close to the poles on each revolution of the earth. They are said to be "sun-synchronous," which means that they are able to view any place on earth and will view each location twice a day at a time near local solar time. Because they are closer to the earth than the GOES satellites, their images have a better resolution and are used to monitor many things ranging from tropical storms to forest fires. Great improvements over time have been introduced and the United States currently uses the satellites designated by "NOAA" followed by a letter of the alphabet which changes to a number after successful launch. In 2007, NOAA 17 and NOAA 18 were the primary NOAA satellites.

Satellites carry a wide variety of sensing instruments, but essentially may be classed as imaging systems and sounding systems. Imaging systems are either (a) those that give a series of instantaneous images, or (b) scanning sensors that build up images line by line in tracks. Sounding systems are those that measure emissions of radiation from different levels of the atmosphere to allow the construction of atmospheric profiles.

FIGURE 6.2. The National Weather Service Office in Indianapolis, Indiana.

can be used to prevent local forecasters from reinventing the wheel. However, material from national centers needs modification before it can be applied to an area the size of Indiana. Local features such as Lake Michigan, the hills in the southern part of the Hoosier State, or even the flood plain of the Wabash River, produce effects a national product cannot capture. Some flaws are also unavoidable in national products if they are to be produced in a reasonable amount of time. These are usually negligible on the large scale, but significant improvement may be possible by an expert on a smaller scale.

Forecasting at the local level is a process of deciding what material will be useful from the national centers, making adjustments for any local errors in those materials, and allowing for effects of terrain close to the forecast area. Weather prediction is like medical diagnosis—some things may be known directly, but much depends on the experience and skill of the practitioner. Although basic forecast theory is usually learned in a four-year college program, true proficiency requires years of practice beyond that.

In an FO, responsibilities tend to be divided according to who ultimately *receives* a forecast product. This is because although the basic information may be similar, different needs determine how forecast products are structured, and hence how they are prepared. For instance, the most important product released to the public at large is the "gridded" forecast. The service area for a given FO will contain thousands of grid points, all stored in the AWIPS computers. Each point represents a different location and is associated with numbers representing different forecast elements at different times. For example, a particular grid point will have the high and low temperature for each of the next seven days, as well as much other information associated with it. It is the job of the forecaster to ensure that these numbers are the most accurate possible—with the help of software, of course. A person who has the proper computer resources can interrogate a grid point through

FIGURE 6.3. The forecast area of the NWS Office in Indianapolis, Indiana.

the internet to find exactly what is being forecast at a precise spot. Gridded forecasts have revolutionized how meteorological information is conveyed.

More specialized activities are also considered part of public services. For instance, forecasts are issued to support wildfire control. These also use grids, but they have to contain special elements such as smoke dispersal. Another task that is part of public forecasting is issuing long-term (more than six hours in the future) watches, warnings, and advisories for events such as snow storms. These can be related to gridded forecasts, but pose a distinct problem because they are sporadic but very important.

No forecast is helpful unless it is accurate, and people often wonder about the accuracy of the NWS. Verification of forecasts is a surprisingly complex issue depending on what you want to measure, and it cannot be adequately discussed in a short space. However, NWS forecasts are routinely compared to corresponding observations and the results are given to forecasters. For temperatures, most predictions by the FO at Indianapolis out to three days are accurate to within 3.5°F.

Aviation Services

Everyone gets the same weather, but not everyone cares about the same aspects of it. Pilots and others in the aviation industry need information about clouds, visibility, and exact weather arrival times that is not usually critical to other people. To meet these needs, FOs have aviation-specific services. The most important of these is the Terminal Aerodrome Forecast (TAF). TAFs are issued four times a day. They contain detailed information on aviation weather for the next twenty-four hours—for instance, specific predictions of cloud amount, cloud height, and surface visibility. TAFs also contain very specific forecasts about the direction and speed of the wind, and any wind gusts. If appropriate, TAFs mention other types of weather that affect air travel, such as thunderstorms or freezing rain. Since

```
000
FTUS43 KIND 011700
TAFLAF
TAF
KLAF  011725Z 011818 16003KT P6SM  SCT080
      FM0900 00000KT  4SM  BR  SKC
      FM1300 22004KT P6SM  SKC=
```

FIGURE 6.4. TAF for Lafayette Airport. Explanation: The 011818 group indicates
a forecast on the first of the month for 18 hours beginning at 18 hours—or 6PM—
Greenwich Mean Time (2PM EDT). The initial forecast is for winds from 160 degrees
at 3 knots (16003KT), visibilities above 6 statue miles (P6SM), and scattered clouds at
8 thousand feet (SCT080). At 5AM EDT (indicated by FM0900), the wind is forecast to
be calm (00000KT), visibilities will be 4 miles in fog (4SM BR), and skies will be clear
(SKC). By 9AM EDT (FM1300), the wind is predicted to be from 220 degrees at 4 knots
(22004KT), visibilities will be above 6 miles (P6SM), and skies should remain clear
(SKC=). Figure courtesy of John Kwiatkowski.

the aviation industry is very weather sensitive, TAFs are constantly monitored and are
updated for only slight deviations between what is expected and what is observed. Figure
6.4 shows a sample TAF for Lafayette Airport.

As of July 2007, TAFs were produced for four cities in central Indiana: Bloomington,
Indianapolis, Lafayette, and Terre Haute, a typical FO workload. Since they are produced
in relatively small numbers, TAFs have not been automated to the same extent as the public forecasts discussed above. Rather than being developed from a grid, they are entered
manually into a formatting program. However, the number of TAF locations is constantly
increasing and TAFs may be produced from grids in the future.

Like most FOs, the weather forecast office in Indianapolis traditionally had separate
forecasters for aviation and public duties continuously on duty. As technology has evolved
this distinction has broken down. Now it is common to speak of a short term forecaster
and a long term forecaster. The short term forecaster does most things with a time horizon
of less than about twenty-four hours, which includes the aviation products. He or she may
have many other duties such as updating the gridded forecasts as needed through the next
twenty-four hours, and preparing so-called Nowcasts. A Nowcast is a very short-term (usually two hours or less) forecast based largely on trends revealed by radar, satellite, or other
data collection methods. They are particularly important when the weather is changing
rapidly. The long term forecaster concentrates mainly on forecasts for the next two to seven
days, as well as on any long-term products as discussed above.

Severe Weather
When the NWS refers to severe weather, it means specifically tornadoes or thunderstorms
that produce damaging winds or hail. Other types of weather, such as blizzards or flash
floods, can be every bit as important, but are classified differently. Under the above definition, severe weather occurs in central Indiana approximately thirty-five times per year
(there is great variation between years). Safeguarding the population from this is the
highest priority of an FO.

Large-scale severe weather is normally preceded by a watch from the Storm Prediction
Center. Minor severe weather may not be. A watch indicates that there is a significant risk

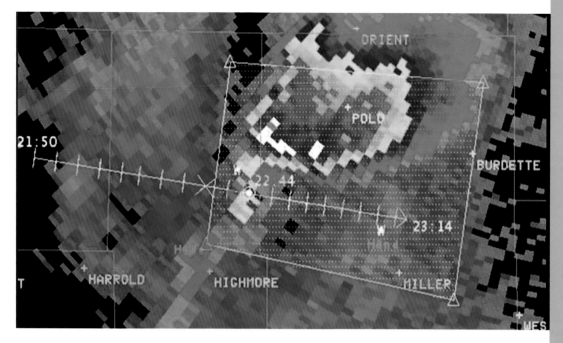

FIGURE 6.5. A polygon warning being formulated. The area in the box corresponds to the greatest threat. Figure courtesy of John Kwiatkowski.

of severe thunderstorms or tornadoes in the near future. FOs help disseminate watches and are responsible for clearing them after danger passes. However, the primary job of an FO during severe weather is to issue warnings. Warnings indicate severe weather is either in progress or about to occur. Immediate action is necessary to ensure your safety. Warnings are issued in the form of polygons indicating the area a storm will affect over the next hour or so. Depending on storm characteristics, a polygon could cover part of just one county, or it might span several. A polygon warning being prepared is shown in figure 6.5.

An FO uses many sources of data in deciding whether to issue a warning, but the most important is radar. Doppler radars such as the NWS WSR-88D gather a wealth of information about both storm shape and motion within a storm. Looking at both motion and shape is much more powerful than seeing either alone. For instance, a storm with the famous hook echo appearance (figure 6.5) would immediately be a candidate to produce a tornado (see chapter 2). But not every storm that is a possible tornado producer displays the hook echo form. Seeing the motion *within* a storm can often reveal relatively subtle information about severe storms. Besides radar information, clues to whether a storm is becoming tornadic can come from what storm spotters see and the environment in which the storm developed. Making a warning decision is like solving a crime—bits of evidence are pieced together until a coherent picture emerges.

Given a decision to issue a warning, special software on the AWIPS computer is used. This enables the forecaster to specify such parameters as warning type (tornado or severe thunderstorm), storm speed and direction, warning length, etc., in seconds. The newly created severe thunderstorm or tornado warning is then sent to the world at large. It will be followed up by periodic severe weather statements, which update information on the storm and discontinue the warning where it is no longer needed.

Oh, the wonder and glamour of being a morning television weathercaster. I awake at 1:50AM and even while in the shower start thinking of the day ahead and what the top local weather story will be. After watching the 11PM newscast (played back at 2AM on WTHR-TV), I am out the door, in my car, and on my way for the twenty-minute trip downtown to Indy.

At 2:50 AM I meet with producer Cyndee Hebert. We discuss the weather outlook for the morning rush hour and we also discuss other components of the show, like major news stories, sports, and the location of our live remote trucks. This is important because we use a more holistic approach to the news on our Sunrise broadcast. I will be on the set talking about weather, of course, but also in on discussions on subjects ranging from a national election to the Indianapolis Colts' latest victory. Now it gets intense. Since I am on the air on television at 4:30AM and radio at 4:20AM—I prepare and voice forecasts for three radio stations—I must really move and groove in preparing both an accurate forecast and, of course, visually compelling graphics for the weathercast.

I will review our forecast from the night before and then take a look at overnight computer models to make the tweaks or adjustments needed to update the forecast. I really like three main computer models. The surface map shows the position of warm and cold fronts, areas of low and high pressure, and differences in barometric pressure. That serves as a cursory look at how the weather may change. The 500mb—or millibar (a unit of pressure)—chart shows me the general direction of the weather by identifying contours in

Warnings are disseminated in a variety of ways. Most people receive warnings through television in a joint effort of the NWS and broadcast meteorologists. Radio and various internet services are also very important for warning distribution. Sirens are useful in some areas, but it is hard to guarantee that they will be audible during a storm. They should not be the first line of defense in severe weather.

Since severe weather *must* be dealt with immediately, a major event dramatically increases the pace at an FO. Extra staff will likely be called in. Different individuals may be assigned to issue warnings for different parts of the service area. Other people will work with storm spotters, disseminate information, and do analysis to provide meteorologists with supplemental environmental information. A large severe weather outbreak could see ten or more people (the normal staffing is three people) working eight hours or so. However, an event so extreme occurs only a few times a decade. Most severe weather episodes in central Indiana are comparatively small. But whether a round of severe weather is large or small, the NWS must be ready to meet it.

Hydrology and Other Vital Tasks
In addition to HMTs and meteorologists, many offices have a Service Hydrologist (SH). The most important thing this person does is to forecast river stages based on rainfall. This is important in determining whether flooding will occur, and if so, to what extent—a truly vital task as flood plain development continues in many areas. Of course, in most locations flooding is only a problem for a small portion of the year. However, the job of the SH does not end—his or her forecasts are still crucial in effective water management. Besides forecasting water levels, the SH is active in maintaining the data collection system along watercourses. He or she frequently visits stream gages to keep them up to par and is also involved in maintaining the electronic systems that collect and transmit river data.

An aspect of FO operations that not many people know about is the National Oceanic and Atmospheric Administration Weather Radio (NWR) network. Most FOs have radio stations that broadcast weather information twenty-four hours a day; all of Indiana is within range of such stations. These stations are largely automated. Most information produced by an FO is passed to the airways without manual intervention. This means warnings can be transmitted on NWR within seconds of composition. Modestly priced receivers can be programmed to sound an alarm **only** when warnings are issued for a particularly county. This feature plus the speed by which materials can be transmitted make NWR the best way to receive severe weather information for most Indiana residents

The Cooperative Weather Program

Roger Kenyon

Every day, thousands of volunteer weather observers record twenty-four-hour rainfall amounts, as well as maximum and minimum temperatures, as part of a program of the NWS. These volunteers are cooperators in a network of weather observation sites called the Cooperative Weather Observer Network (COOP) which comprise the backbone of the climatological data base for the United States and its territories. More than 11,000 volunteers take observations on farms, in urban and suburban areas, National Parks, seashores, and mountaintops. The data are truly representative of where people live, work, and play.

In Indiana, there are well over two hundred COOP stations. The network is the most cost-effective program in the United States government since few volunteers are paid. Government funds are used for equipment and maintenance of the program. The COOP program is truly the nation's weather and climate observing network of, by, and for the people. The COOP program's mission is two fold.

the air above our heads. Although the jet stream is an important tool to use, and most television weathercasters use it, the 500mb chart contours help me decide in which directions the systems are moving. The 850mb chart looks a lot like the 500mb chart but is lower to the ground. This helps me predict when rain or snow may occur.

After putting together a rough forecast, I check data provided to us from our vendor Weather Central and the NWS. Hourly weather graphs, moving surface map models, radar, satellite, and current observations around the Midwest all go into my preparation for a forecast. I then decide the story of the day. Is it a rainy start or a snowy afternoon . . . a minor heat wave or a cold, arctic blast? Once I decide the story of the day, I prepare my weather graphics on our computer on that idea and base my presentation from that single thought. Since I do eleven separate weather forecasts during the sunrise broadcast and eight more during the Today show, I have to vary my presentation somewhat yet still keep the story of the day as the main focus. I also am constantly on the air on three radio stations both informing people of the weather on their way to work and making sure I reinforce that we will have more detailed information on WTHR-TV. It is a great job. I have had the pleasure of keeping people safe in severe weather and encouraging them to go outside to enjoy the weather on a beautiful day.

1) To monitor observational meteorological data, usually consisting of daily maximum and minimum temperatures, snowfall, and twenty-four-hour precipitation totals, all of which are required to define the climate of the United States and to help measure long-term climate changes.

2) To provide nearly real-time observational meteorological data to support forecasts, warnings, and other public service programs of the NWS.

Here at the outset of the twenty-first century we are accustomed to having technical information at our fingertips. This is also true with weather data. The advent of high speed computers, satellite imagery, and, more recently, Doppler radar, has given the weather studying community advantages meteorologists and climatologists only dreamed about one or two decades ago. Although the means by which data is available to the general public has improved tremendously over the past decade or two, the actual weather observation—for climate purposes, for example, the temperature variance and the amount of precipitation—is obtained much as it was one hundred or even two hundred years ago.

History

We need not look back very far to recognize the beginnings of this all-important data activity. The unheralded pioneers of the study of meteorology and climatology are many, and the same spirit that encouraged those of yesteryear to observe the elements continues in many Americans today. The first man to take systematic weather observations in the New World was a minister named John Campanius Holm. This was done in 1644–1645, without instruments, in Swedes's Fort which is now Wilmington, Delaware. The weather observation road from Holm's time through the mid-1800s is rocky, but interesting. Benjamin Franklin, George Washington, James Madison, and Thomas Jefferson were all weather aficionados. Franklin is known to have been the first to track a hurricane using local postmasters as observers, while Jefferson and Madison took the first simultaneous weather observations. Jefferson himself logged an almost unbroken account of observations for some forty years, while George Washington took his last weather observation just a few days before he died.

In Indiana, weather records of the 1800s are not numerous. Weather records were often kept in diaries or with some other method of recording data and thus were mostly lost. Also at that time very few weather stations existed in Indiana, and communications were poor.

There was no known statewide summarization of data on a monthly basis until 1882. At least two networks, including the Smithsonian and the Surgeon General's Office, existed in Indiana where, to some extent, temperature and rainfall data were duplicated.

One of the oldest locations for weather observations in the Hoosier State is New Harmony whose records date back to 1826 when observations were taken from three to nine times daily. The first well-organized and lengthy records were recorded by the settlers in New Harmony between 1853 and 1883. Other locations taking observations during the mid-nineteenth century include La Porte, which began in 1849, Logansport in 1854, Richmond in 1852, South Bend in 1862, and both Collegeville and Vevay in 1864.

The number of weather observers in Indiana increased rapidly beginning in the 1860s.

The Cooperative Observer

Volunteer weather observers conscientiously contribute their time so that observations can provide the vital information needed. These data are invaluable in learning more about the floods, droughts, as well as the heat and cold waves affecting the nation. The data are also used in agricultural and utilities planning, engineering, environmental impact assessment, and litigation. COOP data play a critical role in efforts to recognize and evaluate the extent of human impacts on climate from local to global scales.

The bulk of the cooperative program is made up of outstanding citizens who have a strong interest in the weather. These individuals vary in professions from bankers and salesmen to homemakers and farmers. The NWS has an awards program for these valuable people. Generally, the Length of Service Award is given for every five years of service. Many institutions, such as universities, waste water treatment plants, water companies, and other utility companies, record daily readings. Institutional awards are presented for every twenty-five years of service.

The Thomas Jefferson Award is given to five individuals nationally per year, and the John Campanius Holm award is given to twenty-five individuals per year, to those who show extraordinary expertise for extended periods of time. Several cooperative observers in Indiana have been recipients of one of more of these awards in the last two decades.

FIGURE 6.6. The Jefferson Award is presented to Cooperative observer Warren Baird who is shown on the right.

The United States Historical Climatology Network (USHCN) database represents the best monthly temperature and precipitation dataset available for the contiguous United States. It provides an accurate, serially complete, modern historical climate record that is suitable for detecting and monitoring long-term climatic changes on a regional scale and may be used for studies attempting to determine the climatic impacts of increased concentrations of greenhouse gases.

Only those stations that were not believed to be influenced to any substantial degree by artificial changes of local environments were included in the network. Some of the stations in the USHCN are first-order weather stations, but most were selected from cooperative weather stations. To be included in the USHCN, a station had to be active (in 1987), have at least eighty years of mean monthly temperature and total monthly precipitation data, and have experienced few station changes. An additional criterion that was used in selecting the 1221 USHCN stations, which sometimes compromised the preceding criteria, was the desire to have a uniform distribution of stations across the United States.

The 1221-station USHCN database contains station histories, monthly temperature (maximum, minimum, and mean) data, and total monthly precipitation data that were compiled by NCDC after being extracted from digital and non-digital datasets archived at NCDC. These datasets originated from a variety of sources, including climatological publications, universities, federal agencies, individuals, and data archives. All stations were quality controlled by NCDC, and each

Station Requirements

Proper instrumentation, exposure of instruments to the elements, observation times, and punctilious observers are paramount in the data collection that informs our knowledge of Indiana's climate. At each site rainfall and maximum and minimum temperatures are recorded daily, 365 days a year. "Rainfall only" stations (additional sites based on watershed requirements) are added to supplement the climate data and to supply hydrologic data for river and stream forecasts. The data are collected both on forms and electronically at local NWS FOs where it is checked for quality. It is then sent to the National Climatic Data Center (NCDC) where it is digitized, archived, and published. The computers at NCDC compile extremes, averages, and normals from these data which are then made available to the public.

Instrumentation

Each site must have an NWS-approved temperature shelter and thermometer or thermocouple as well as an eight-inch standard rain gage. During the late 1980s the NWS adopted an electronic temperature device known as the Maximum-Minimum Temperature System to replace the "liquid in glass" thermometers. It is now the more common temperature equipment in use.

Installation

Each installation must be accomplished by an NWS representative. This is to ensure that all the required parameters are being met. Subsequent annual visits are made to maintain integrity.

Exposure to the Elements

Requirements must be maintained for distance and elevation of surrounding trees, buildings, lakes, ponds, or anything that may cause data to be skewed as a consequence of poor exposure. Instruments should remain in the same location as long as possible.

Time of Observation
It is critical that all observers record observations at the same time each day.

Missing Data
If a site has excessive missing data, or if other parameters are not being met and corrected within a reasonable amount time, corrective action must be taken or it must be moved to a location within five miles and one hundred feet elevation.

Acquisition of Data

Many observations are forwarded to the NWS by an automated phone system or on the internet. All observations are recorded

station in the network was corrected for time-of-observation differences, instrument changes, instrument moves, station relocations, and urbanization effects.

An in-depth study of Indiana USHCN stations was completed by Ashley Victoria Brooks of the Indiana State Climate Office at Purdue University. She examined all of the weather stations and ranked them according to the location and exposure of instruments. The rankings (from good to poor) showed that Indiana sites vary appreciably. A good site is illustrated in figure 6.8.

FIGURE 6.7. A well-equipped weather station that is part of the Cooperative network in Indiana.

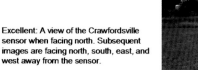

Excellent: A view of the Crawfordsville sensor when facing north. Subsequent images are facing north, south, east, and west away from the sensor.

FIGURE 6.8. The views from a well-situated weather station that forms part of the U.S. Historical Climatology Network in Indiana. Photos courtesy of Ashley Victoria Brooks.

on government forms for publication. Stations in the cooperative program follow distinct guidelines. There are networks that comprise the gathering of data for *climatology* and *hydrology* as well as *supplemental* needs. The climatology network is the largest and includes those stations that record daily maximum and minimum temperatures and precipitation levels. The *hydrology* network is composed of rainfall, river, evaporation, and hourly precipitation recorders (often known as the Fischer & Porter network). The *supplementary* network is made up of stations that provide agriculture data and also includes those who want to be involved and are willing to provide their own NWS-approved equipment.

Data are sent to the NCDC in Asheville, North Carolina, where computers sift through the cooperative data carefully to see if there are errors. From NCDC the data are made available online as well as in published form.

7

INDIANA'S PRESENT CLIMATE—
THE LARGER VIEW

World climates are extremely diverse. Some places, like the equatorial lands, are always hot; some, such as the polar regions, are always cold; and some, like Indiana, experience alternating hot and cold seasons. At the same time, precipitation is not evenly distributed over the globe. The great world deserts experience perpetual drought while other places, like the windward mountains of Hawaii, have many hundreds of inches of rain each year. Indiana is fortunate in that, for the most part, it receives enough monthly rainfall to sustain its water supplies and enable productive agriculture.

To understand the diversity that exists, climatologists have produced classification schemes that group similar climates. Using such schemes, climates like those in Indiana can be identified in various parts of the world. Some areas of Central Europe and Japan are often grouped with the Midwest climate. In the Southern Hemisphere, the lack of large continents means that few places experience the hot summer and cold winter regime of our region.

The study of world climates goes beyond merely identifying types of climate and attempts to understand not only the normal climate, but the unusual as well. In this regard climatologists look to teleconnections.

Connecting Climates

The formidable word "teleconnections" describes nothing more than relating one event in the world to another that is remote from it. In relation to climate, the best known teleconnection is El Niño, an ocean current off South America that influences climates in North America and other places; this event requires special attention and is dealt with a little later. But El Niño is but one teleconnection that deserves our study. Among the many others is the Pacific–North American Oscillation, the PNA. As its name implies, it deals with patterns over the northern Pacific Ocean and the North American continent and the way that the patterns change, or oscillate, season to season and year to year.

The thermocline is a layer within a body of water where the temperature changes rapidly with depth. Because water is not perfectly transparent, most sunlight is absorbed in the surface layer. Wind and waves circulate the surface water distributing heat within it, and the temperature may be quite uniform for the first few hundred feet. Below this mixed layer, however, the temperature drops very rapidly—perhaps as much as 36°F (20°C) with an additional 500 feet (150 m) of depth. This area of rapid transition is the thermocline. Below the thermocline, the temperature continues to drop with depth, but far more gradually.

The PNA is actually an alternating pressure pattern that occurs over the Pacific Ocean, western Canada, and southeastern United States. The pressure patterns occur in winter and sometime lead to the unusual weather conditions that occur in the United States. The PNA is studied as part of the waves associated with the upper air flows described earlier in terms of the mean position and character (for example, the presence or absence of large south-north meanders) of the jet stream. Of the many studies relating the PNA to North American climate anomalies, of particular interest is one that deals with the Midwest. The study showed that precipitation patterns in southern Indiana can be related to variations in the PNA; additionally, in identified PNA conditions, the winter stream flow in the Ohio River was shown to be greatly influenced by the PNA. Identifying teleconnections and their influences is an important component explaining climates in a modern context. Needless to say, some of the work is somewhat complex and deals with statistical models.

Of the many teleconnections that lead to climate differences far from the initial source, El Niño is by far the best known. This feature is associated with changing pressure patterns; in this case, a pressure pattern over the southern Pacific Ocean called the Southern Oscillation. They are often studied together and referred to as El Niño–Southern Oscillation (ENSO) events.

El Niño and La Niña

The El Niño–Southern Oscillation, the most far-reaching global teleconnection, is the periodic shift in ocean and wind patterns in the central Pacific Ocean basin. ENSO events cause dramatic changes in the strength and direction of the trade winds, in sea surface temperatures, and in the depth of the oceanic thermocline.

El Niño, the "Christ child," first named by Spanish colonists for its onset near

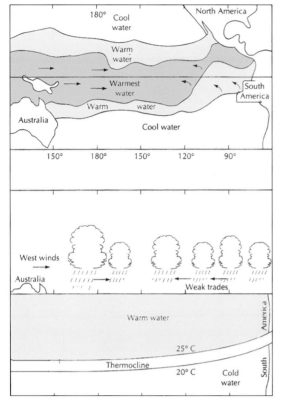

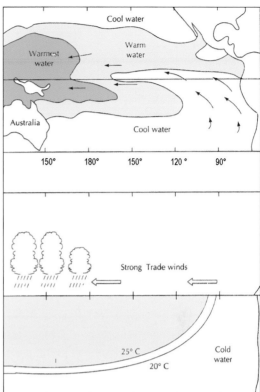

FIGURE 7.1. Conditions that occur in an El Niño phase. The trade winds are weakened and warm water stretches across much of the Pacific Ocean (upper panel). A cross-section shows a modified thermocline location and precipitation in the center and eastern parts of the Pacific Ocean (lower panel).

FIGURE 7.2. Conditions that occur in a La Niña phase. Cold upwelling water occurs off the coast of South America and warmest water is in the western Pacific Ocean (upper panel). A cross-section shows the strong trade winds, rainfall in the western Pacific, and the location of the thermocline (lower panel).

Christmas in early winter, is the warm phase of the oscillation. During El Niño, the cold surface current and natural upwelling west of Peru are replaced by the eastward shift of warmer tropical water along the equator. As the pool of warm water grows and is diverted eastward, the depth of the east central Pacific thermocline subsides and subtropical high pressure weakens in response to increased convective activity.

As atmospheric pressure falls above the warm pool, normal circulation (called the Walker circulation), which typically drives trade winds westward from the east Pacific high toward regions of normally lower pressure over Australia and Indonesia, weakens or even reverses. As the trades slow or become westerly, the eastward movement of warm equatorial waters is enhanced, further strengthening the El Niño episode (figure 7.1).

La Niña episodes are mirror-image anomalies to El Niño in the ENSO cycle of the east central Pacific. La Niñas are characterized by heightened easterly motion in the trade winds, which speeds the southerly cold water current and strengthens the coastal upwelling west of South America, cooling surface waters there and raising the thermocline. Anomalously cold waters in the east Pacific further strengthen the subtropical high, which then more rapidly propel the normal east-west motion of the Walker circulation (figure 7.2).

For purposes of study, ENSO anomalies are defined in multiple ways. One common method for detecting and quantifying the strength of an El Niño or La Niña is to measure the difference (positive or negative) of sea surface temperatures from normal values in the central and eastern Pacific Ocean. For example, the Niño 3.4 is based upon the sea surface temperature anomaly between latitudes 5°N and 5°S and longitudes 170°W and 120°W. Other commonly used measures sample the features of the Walker circulation directly. One such circulation measure, the Southern Oscillation Index, calculates the pressure difference between Tahiti and Darwin, Australia; reversals in this gradient signal an El Niño. Although some debate exists about which index should be used, descriptions of the relationship between ENSO and the climate of a region are often based on a composite of multiple metrics and studies. In the text below, El Niño events are warm sea surface temperature anomalies with their attendant circulation changes, while La Niña events are cold phase anomalies; all other scenarios are ENSO-neutral, that is, where no significant anomaly is present.

For both El Niño (warm) and La Niña (cold) episodes, the frequency of events is variable, typically one event per two to seven years. Further complicating the climatology of ENSO events is the fact that events seldom peak at the same time of year or season. Anomalies are quite irregular in their severity and duration, and also in the duration of the transition period between anomalies. Thus, when describing the influence of ENSO events on Indiana weather and climate, not all El Niños or La Niñas will have the same effect.

ENSO events teleconnect weather changes to the mid-latitudes and Indiana by repositioning and reshaping the jet stream and wave features in the Westerlies. During the warm El Niño, the jet stream divides into two major corridors for storms. The southern, subtropical jet is the dominant mode of this pattern. Heightened energy along the southern corridor tends to favor cyclonic storm formation on the Gulf of Mexico and Atlantic Ocean coastlines. The weaker northern jet stream flows across the northern United States and dips south of the Great Lakes as it enters the Midwest. The moderate northern stream diverts mild Pacific Ocean air masses into the Great Plains and advects cool air toward the central United States. The split-flow pattern that produces the northern and southern jet streams is sometimes associated with a condition known on the west coast as blocking (figure 7.3).

Conversely, cold phase La Niñas are characterized by a single dominant jet core, which enters the United States from the Pacific Northwest and then shifts southward over the Great Plains. The upper-level wave pattern during La Niña positions a strong ridge west of Indiana and delivers cool, dry air into the Great Plains and Midwest.

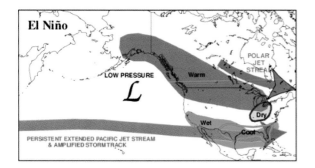

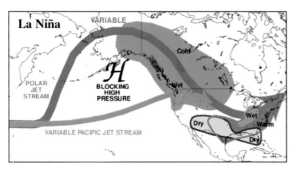

FIGURE 7.3. Schematic diagram showing how El Niño (upper panel) and La Niña (lower panel) conditions influence circulation over North America. Figure courtesy of the National Oceanic and Atmospheric Administration (NOAA) and the National Aeronautics and Space Administration (NASA).

ENSO and Indiana Climate

Greg Bierly

Although the influence of El Niño on certain regions of the world is well known and well documented—rains and flooding in northwestern South America, severe drought in Australia and Indonesia—the effects of ENSO in the United States, and particularly Indiana, are subtle and variable. The following section describes how El Niño and La Niña impact key aspects of Indiana climate.

Temperature

During winter, the occurrence of El Niño and its split-flow and weak northern jet stream result in slightly warmer than normal temperatures in Indiana. This temperature increase may approach 1.8°F (1°C) where weak northwesterly flow and diminished cold air transfer supplant the normal pattern during El Niño in the northern reaches of the state (figure 7.4a). Average temperatures are warmer but nonetheless near normal in southern Indiana. La Niña's adjustment to average winter temperature in Indiana is not symmetrical with, nor opposed to, that of the El Niño events. Most studies indicate that La Niña winters experience near normal (or slightly above normal) temperatures across Indiana (figure 7.4b).

For both phases of the ENSO cycle, the range of temperatures and their distribution around mean values are altered during winter. El Niño events tend to moderate winter temperatures, reducing extremes by diminishing within-season variance around the mean. La Niña winters, on the other hand, tend to be of near average temperatures but may be characterized by high and low temperature extremes.

If anomalous ENSO patterns carry over to the summer months, the influence on mean Indiana surface air temperature becomes less noticeable. When the warm phase El Niño is present, Indiana summer temperatures tend to be near or slightly above normal.

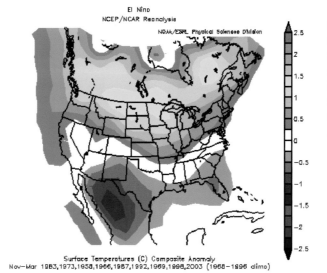

FIGURE 7.4A. Surface temperature patterns for the colder half of the year (November–March) with El Niño circulation. Figure courtesy of the National Oceanic and Atmospheric Administration (NOAA).

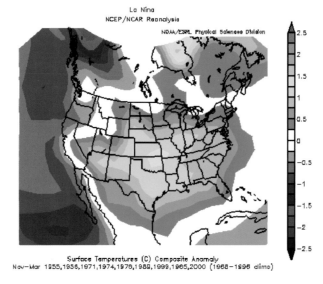

FIGURE 7.4B. Surface temperature patterns for the colder half of the year (November–March) with La Niña circulation. Figure courtesy of the National Oceanic and Atmospheric Administration (NOAA).

As with winter temperature patterns, the changes during El Niño events are greatest in the northern sections of the state (figure 7.5a). During summer, La Niña episodes are associated with temperatures slightly above normal across most of Indiana (figure 7.5b).

Precipitation

For the last half of the twentieth century, winter precipitation in Indiana has been typically near normal over the southern two-thirds of the state and slightly above normal (100–120% of normal) over the northern third of the state during particularly strong El Niños. As the strength of the El Niño is removed, and all events are considered, its relationship to Indiana precipitation blurs and possibly reverses. However, during spring (January–May) following strong El Niños, Indiana precipitation is below normal (60–70% of normal). La Niña appears to reduce winter precipitation in Indiana, perhaps by as much as 2–3 inches over the entire season.

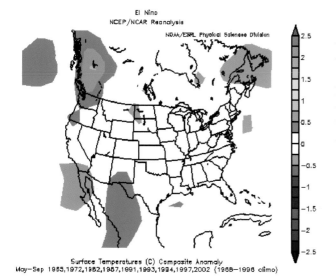

FIGURE 7.5A. Surface temperature patterns for the warmer half of the year (May–September) with El Niño circulation. Figure courtesy of the National Oceanic and Atmospheric Administration (NOAA).

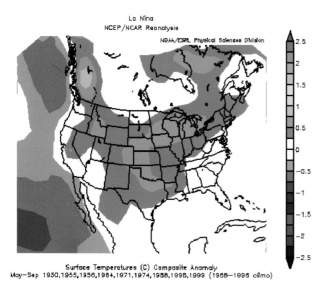

FIGURE 7.5B. Surface temperature patterns for the warmer half of the year (May–September) with La Niña Circulation. Figure courtesy of the National Oceanic and Atmospheric Administration (NOAA).

Both El Niño and La Niña conditions increase summer rainfall totals in Indiana, particularly La Niña. During an El Niño year, summer precipitation rises by as much as 2.5 inches across the northwestern section of the state, and by 1–2 inches in the south. La Niña years see summer increases of 3–4 inches of rain in most areas of Indiana. Neither ENSO phase is associated with a significant change in dynamic vertical motions during summer, so it may be surmised that precipitation increases are connected to enhanced convective energy and available moisture.

In addition to modifying the average summer precipitation levels, climatologists have suggested that the frequency and intensity of rainfall episodes in the Midwest also are modified by ENSO; this has implications for drought establishment, severity, and persistence. For instance, summer rainfall totals during El Niño periods tend to be produced by more constant and moderate rainfall events than those of La Niña; the cold phase anomaly is associated with episodic and often more intense events.

Snowfall

There is general agreement that El Niño events decrease Indiana snowfall; the link between La Niña and snowfall is less clear for the winter considered as a whole, but if winter is subdivided into early, middle, and late periods, it appears to reduce snowfall during these periods. Smith and O'Brien (2001) in particular found significant relationships between cold phase events and snowfall.

In the early winter months (October–December), Indiana snowfall totals are lessened during both warm and cold ENSO phases of the anomaly in comparison years characterized by neutral conditions. For both anomalies, this relationship weakens or disappears in the extreme southwestern portion of the state. Midwinter in Indiana (December–February) also sees reductions in snowfall totals with both El Niño and La Niña relative to neutral conditions; again, this connection is most tenuous in the south. Snowfall amounts in the northern zones during ENSO anomalies are approximately 25 percent of their typical midwinter total. During late winter, no coherent ENSO phase/snowfall relationships are apparent for the Midwest or Indiana.

Severe Weather

Because ENSO events influence the orientation and amplitude of mid-latitude waves of the Westerlies, it is likely that some relationship exists between Pacific sea surface temperature anomalies and severe weather in the Midwest. However, such relationships are hard to quantify and much more research has been devoted to the effects of ENSO events on planetary or regional scale phenomenon than on mesoscale and microscale features.

Since 1998 a number of analyses by various researchers seem to confirm a relationship between ENSO events and midwestern tornadoes. El Niño generally produces weaker tornadoes (fewer F4 and F5 tornadoes) and is less likely to generate large outbreaks of tornadoes (forty or more) than neutral or La Niña conditions. This reduction in tornado activity during El Niño is particularly significant in the southern Great Plains states and less demonstrably so in the upper Midwest/Great Lakes region.

Indiana tornado activity appears to be most influenced by La Niña episodes. La Niña years see an increased number of tornadoes and large tornado outbreaks in much of the Midwest, with numbers nearly doubled from El Niño and neutral years. Devastating F4 and F5 tornadoes are also more frequent during La Niña. These effects are most prominent during early and mid-spring and diminish as summer approaches. When the Midwest study region is partitioned by grid areas, the zone nearest Indianapolis experiences some of the most dramatic La Niña–related tornado effects of the entire state and region. As Bove (1998) notes, the probability of four or more tornadoes near Indianapolis during early spring increases by 300 percent during La Niña years (as compared to neutral years). Probabilities for the occurrence of five or more tornadoes in the vicinity of Indianapolis increase by 600 percent during La Niña. As the size of the outbreak increases, the percent disparity between neutral and La Niña years continues to rise, until the size of the outbreak becomes so large that it never appears during neutral years.

The study of the influence of ENSO events, like that of any other atmospheric phenomena, requires climatic data. It is most fortunate that there are data centers that are able to provide much of what is needed. That applicable to Indiana is now examined.

The Midwestern Regional Climate Center

Kenneth E. Kunkel

The National Oceanic and Atmospheric Administration's (NOAA) Regional Climate Centers are a federal-state cooperative effort formed as a result of the Climate Program Act of 1978. Six existing centers, located at universities and state agencies in Illinois, New York, Louisiana, Nebraska, Nevada, and North Carolina provide quality data stewardship, improved use and dissemination of climate data, and information for the economic and societal good of the United States. The centers also conduct applied climate research in support of improved use of climate information. One of three regional climate center demonstration projects was located at the Illinois State Water Survey (ISWS)* from 1981 to 1986, and the Midwestern Regional Climate Center (MRCC) was formally established at the ISWS in August 1987. The MRCC region includes the states of Illinois, Indiana, Iowa, Kentucky, Michigan, Minnesota, Missouri, Ohio, and Wisconsin (figure 7.6).

FIGURE 7.6. States associated with each regional climate center. Figure courtesy of the Midwestern Regional Climate Center.

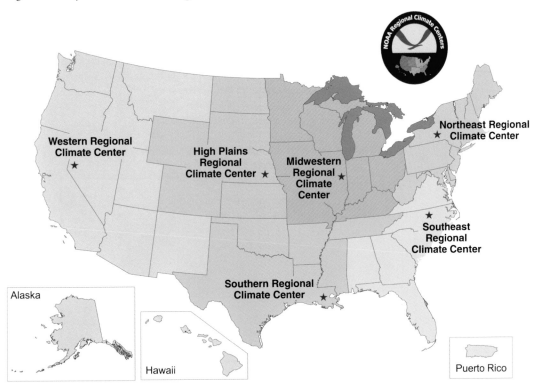

* The Illinois State Water Survey is a division of the Illinois Department of Natural Resources and an affiliated agency of the University of Illinois at Urbana-Champaign.

Climate Data

The activities of the MRCC rely on stations distributed throughout the region that collect high quality climate data and keep long-term records so that current conditions can be evaluated in the context of historical experience. The core climate network of the United States for gathering temperature and precipitation data is the National Weather Service's Cooperative Observer Network (COOP), described in chapter 6. This is a rather dense network (about 11,000 stations are currently active) largely made up of volunteers, both individual and institutional, who take observations once per day. Some stations in this network have records extending back to 1900 or earlier.

The MRCC also gathers data from other networks, most prominently airports where other climate variables, such as pressure, winds, clouds, and weather types are observed.

Climate Monitoring and Information Dissemination

A major MRCC activity is the monitoring of the climate of the Midwest. This region encompasses much of the nation's corn- and soybean-producing areas, much of the United States's portion of the Great Lakes Basin, the upper half of the Mississippi River Basin, and several large metropolitan areas. Up-to-date climate information for the region is available through the MRCC's Midwest Climate Watch (MWCW) webpage (http://mrcc.sws.uiuc.edu/cliwatch/watch.htm). The MWCW provides users with a suite of temperature and

FIGURE 7.7. A map of Midwest soil moisture in the top 72 inches for May 23, 2007. Soil moisture is estimated from a model that uses temperature and precipitation as input. Soil moisture is expressed as a deviation (in percentage) from average conditions. Yellow and red shades indicate below average soil moisture, while green and blue shades indicated wetter than average conditions. Figure courtesy of the Midwestern Regional Climate Center.

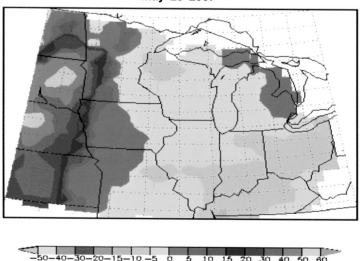

Current Soil Moisture Deviation (%). Depth = 0–72
May–23–2007

Midwestern Regional Climate Center
Illinois State Water Survey
Champaign, Illinois

precipitation maps and other information automatically updated each day. The MWCW page contains many temperature, precipitation, soil moisture, and snow maps of the Midwest region, and links to many other relevant sites. For example, figure 7.7 shows a map of estimated soil moisture relative to normal conditions on May 23, 2007. At this time, a period of dryness had led to soil moisture levels 10 to 15 percent below normal in parts of Illinois, Indiana, Ohio, and Kentucky, while wet conditions were present in the western part of the Corn Belt. Maps like this are updated every day, providing an immediate sense of the evolution of climate conditions.

MRCC staff members compose a weekly narrative of weather and climate conditions for the region. This narrative provides a general description of conditions for the week, highlights significant events and impacts (including charts, photos, and other graphics as appropriate), and provides links to supplementary material. This provides users with a written professional synthesis of the graphical information available in the MWCW, along with other information of interest.

The MRCC website also provides access to the Midwestern Climate Information System (MICIS), the MRCC's online climate data and information system; this has been in operation since 1989. This is a subscription-based system, that is, users pay a small fee to support the infrastructure and equipment needed to maintain the data and product delivery. Subscribers use MICIS to access user-specified climate data and products for a

FIGURE 7.8. Map of temperature for July 1936. Temperature is expressed as a departure (in °F) from the 1971–2000 average. Green, yellow, and red shades indicate above average temperatures, while blue and purples shades indicate below average temperatures. Figure courtesy of the Midwestern Regional Climate Center.

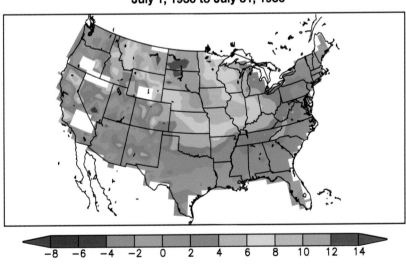

Average Temperature Departure from Mean in Degrees F
July 1, 1936 to July 31, 1936

Midwestern Regional Climate Center
Illinois State Water Survey
Champaign, Illinois

Table 7.1. Precipitation Percentiles for Bloomington, Indiana, 1900–2006

	1%	5%	10%	25%	50%	75%	90%	95%	99%
Ja	0.20	0.53	0.82	1.57	2.86	4.73	6.99	8.61	12.23
Fe	0.43	0.78	1.04	1.59	2.42	3.50	4.70	5.54	7.34
Ma	0.58	1.10	1.50	2.37	3.70	5.47	7.46	8.85	11.87
Ap	0.96	1.53	1.92	2.72	3.84	5.25	6.76	7.79	9.97
Ma	0.83	1.43	1.87	2.79	4.15	5.89	7.81	9.13	11.97
Jn	0.67	1.21	1.61	2.48	3.78	5.47	7.36	8.66	11.48
Ju	0.68	1.20	1.58	2.39	3.59	5.14	6.85	8.04	10.58
Au	0.55	1.04	1.40	2.19	3.40	4.99	6.78	8.02	10.71
Se	0.35	0.75	1.08	1.83	3.05	4.71	6.63	7.99	10.98
Oc	0.25	0.57	0.85	1.50	2.57	4.07	5.83	7.09	9.86
No	0.66	1.12	1.45	2.14	3.15	4.43	5.84	6.80	8.87
De	0.60	1.05	1.38	2.07	3.10	4.42	5.89	6.90	9.08
An	26.49	31.79	34.95	39.12	44.13	49.55	54.79	58.10	64.62
Wi	3.09	4.40	5.24	6.88	9.09	11.73	14.49	16.33	20.16
Sp	5.60	7.25	8.25	10.14	12.56	15.33	18.15	19.99	23.75
Su	5.09	6.62	7.56	9.32	11.59	14.20	16.85	18.58	22.13
Fa	3.09	4.47	5.37	7.12	9.50	12.35	15.35	17.36	21.55

Note: Entries in the leftmost column are (from top to bottom): January through December, annual, winter, spring, summer, and fall. Precipitation totals are given in inches. Precipitation percentiles for Bloomington, Indiana, come from the Midwestern Climate Information System. For example, the 90th percentile value for July is 6.85 inches, which means that precipitation totals will be below that value in nine out of ten Julys. Data recorded at the Daily Historical Climatology Network station at Indiana University, Bloomington (120784); missing data: 2.7%.

variety of applications. Users are able to customize products for a specific location and/or time period, and can produce output in a number of formats.

Examples of MICIS products illustrate the flexibility and breadth of available information. Table 7.1 shows a summary of the precipitation climatology for Bloomington, Indiana. This table gives precipitation amounts for various probabilities (percentiles). For example, in 5 percent of years, July precipitation has been less than 1.20 inches. The 95th percentile value for July is 8.04 inches, which means that this value has not been exceeded in 95 percent of all July's on record. Table 7.2 shows a ranking of summer temperature for the state of Indiana. The warmest summer on record was 1936, while the coolest was 1992. Table 7.3 shows a ranking of summer precipitation for the state of Indiana. This shows that 1936 was also the driest summer on record, while 1958 was the wettest. It is obvious that the summer of 1936 was very extreme. If there is interest in the geographical context of this period, another set of products provides maps of various conditions. Figure 7.8 shows the spatial distribution of temperature anomalies for July 1936. Temperatures averaged 6–8°F above normal in Indiana. The map shows that Indiana was on the eastern side of a very hot area centered on the Dakotas where temperatures were more than 10°F above normal.

The MRCC responds to individual requests for data and information received by phone, mail, and email.

Table 7.2. Ranked State-Averaged Summer (June–August) Temperature (°F) in Indiana, 1895–2006

Rank	Year	Data	Rank	Year	Data	Rank	Year	Data	Rank	Year	Data	Rank	Year	Data
1	1936	76.9	2	1934	76.8	3	1901	75.8	4	1921	75.8	5	1983	75.6
6	1943	75.2	7	1933	75.1	8	1914t	75.1	8	1952t	75.1	10	1944	75.0
11	1913	75.0	12	1995	74.9	13	1988t	74.9	13	2002t	74.9	15	1949	74.9
16	1953	74.8	17	1954	74.8	18	1931	74.8	19	1919t	74.6	19	2005t	74.6
21	1900	74.5	22	1991	74.3	23	1899	74.3	24	1898	74.2	25	1987	74.0
26	1959	74.0	27	1980	74.0	28	1940	74.0	29	1955	73.9	30	1895	73.9
31	1941	73.9	32	1932	73.8	33	1930	73.6	34	1999	73.6	35	1925t	73.5
35	1939t	73.5	37	1973	73.4	38	1937	73.3	39	1916	73.3	40	1938	73.3
41	1942	73.2	42	1918t	73.2	42	1975t	73.2	44	1935	73.2	45	1993	73.1
46	1922	73.0	47	1896	73.0	48	1948	73.0	49	1984	72.9	50	1906	72.9
51	1947	72.9	52	1911t	72.9	52	1968t	72.9	54	1957t	72.8	54	2006t	72.8
56	1927	72.8	57	1923	72.8	58	1905	72.8	59	1908	72.7	60	1998	72.7
61	1956	72.7	62	1977	72.7	63	1981	72.6	64	1909	72.6	65	1986	72.6
66	1964t	72.5	66	1966t	72.5	68	1978	72.4	69	1897	72.3	70	1926	72.2
71	2001	72.1	72	1971	72.0	73	1989	72.0	74	1994	71.9	75	1970	71.9
76	1962t	71.8	76	1969t	71.8	78	1951	71.7	79	1996	71.6	80	1960t	71.5
80	2003t	71.5	82	1902	71.4	83	1946	71.4	84	1974t	71.4	84	2000t	71.4
86	1979	71.3	87	1910t	71.3	87	1990t	71.3	89	1961	71.3	90	1976	71.2
91	1945	71.1	92	1963	71.1	93	1928	71.1	94	1924t	71.0	94	1929t	71.0
96	1912	71.0	97	1965	70.9	98	1907t	70.8	98	1985t	70.8	98	1997t	70.8
101	1920	70.7	102	1917	70.6	103	1972	70.6	104	1904t	70.5	104	1958t	70.5
106	1982	70.4	107	1950t	70.3	107	1967t	70.3	109	1903	70.3	110	2004	70.1
111	1915	69.0	112	1992	68.9									

Note: Ranking is from warmest (1) to coolest (112). The table provides the rank, year, and average temperature ("Data" in °F); "t" means a tie.

Applied Research

One research focus is the study of extreme events. The primary purpose of such studies is to gain a better understanding of the influence of such events on society and the environment. These studies have typically involved both an evaluation of the significance of the climate conditions in the context of the historical record and an assessment of the environmental effects and socio-economic impacts. Some of the notable events that have been studied include:

- 1988 drought: This event was the worst drought and hottest summer in the Midwest since the Dust Bowl era of the 1930s with losses of more than eighty billion dollars (Lamb et al. 1992).
- 1993 Mississippi River flood: This event was triggered by the wettest summer on record in the Upper Mississippi River Basin. Losses are estimated at thirty-four billion dollars (Bhowmik et al. 1994; Kunkel et al. 1994). Figure 7.9 shows the pattern of precipitation during that summer. Precipitation amounts were

Table 7.3. Ranked State-Averaged Summer (June–August) Precipitation (inches) in Indiana, 1895–2006

Rank	Year	Data	Rank	Year	Data	Rank	Year	Data	Rank	Year	Data	Rank	Year	Data
1	1936	6.07	2	1933	6.22	3	1930	6.45	4	1908	6.86	5	1940	7.34
6	1991	7.35	7	1944	7.49	8	1922	7.50	9	1919	7.87	10	1983	7.94
11	1984	8.00	12	1988	8.06	13	1895	8.13	14	1967	8.26	15	1966	8.40
16	1904	8.42	17	1913	8.46	18	1901	8.75	19	1911	8.86	20	2002	8.86
21	1914	8.93	22	1999	9.09	23	1899	9.15	24	1953	9.16	25	1897	9.24
26	1959	9.57	27	1956	9.67	28	1918	9.69	29	1927	9.81	30	1910	9.88
31	1920	9.90	32	1946	10.04	33	1972	10.12	34	1964	10.15	35	1898	10.17
36	1952	10.35	37	1943	10.42	38	1963	10.54	39	1925	10.56	40	1934t	10.57
40	1941t	10.57	42	1974	10.60	43	1955	10.70	44	1948	10.71	45	1965	10.73
46	1970	10.84	47	1976	10.92	48	1917	10.94	49	1903t	11.00	49	1921t	11.00
51	1931	11.08	52	1954	11.13	53	1951	11.21	54	1994	11.22	55	1995	11.25
56	1906	11.28	57	1929	11.34	58	1971	11.38	59	1982	11.55	60	1935	11.58
61	1932	11.74	62	1947t	11.76	62	1962t	11.76	64	1923	11.77	65	1924	11.84
66	1937	11.85	67	1916	11.91	68	1996	11.94	69	1961	12.02	70	1985	12.05
71	1986	12.28	72	1968	12.30	73	1926	12.38	74	2005	12.55	75	1960	12.56
76	1949	12.61	77	1997	12.70	78	1969	12.91	79	1950	12.93	80	1987	12.94
81	1939	12.99	82	1902	13.07	83	1975	13.17	84	1905	13.20	85	1912	13.25
86	1978	13.38	87	1909	13.39	88	1945	13.40	89	1942	13.43	90	1907	13.45
91	1989	13.47	92	1900	13.59	93	1938	13.61	94	1981	13.71	95	2001	13.85
96	1957	14.09	97	2006	14.24	98	1977	14.36	99	1928	14.38	100	1973	14.41
101	2000	14.57	102	1980	14.59	103	1992	14.60	104	1993	14.75	105	2004	14.78
106	1990	14.87	107	2003	15.25	108	1915	15.87	109	1998	16.38	110	1896	17.08
111	1979	18.24	112	1958	20.63									

Note: Ranking is from driest (1) to wettest (112). The table provides the rank, year, and total precipitation ("Data" in inches); "t" means a tie.

more than twice (that is, more than 200 percent of) the normal in parts of Iowa, Nebraska, Kansas, Missouri, Illinois, and near the Canadian border. Precipitation amounts were above normal in Indiana, but only slightly so.

- July 1995 heat wave: A very intense heat wave during July 12–15, 1995, was the third hottest in Chicago history and resulted in more than 700 deaths (Changnon et al. 1996).

- 1997–1998 El Niño: The winter of 1997–1998 was the warmest on record in the Midwest. Although storms accompanying this event caused more than thirteen billion dollars in losses, the winter warmth and lack of snow resulted in gains of more than twenty-four billion dollars, much of this in the Midwest (Changnon et al. 2000). Temperatures were more than 6°F above normal in northern Indiana and more than 10°F above normal in northern Minnesota.

- 2004 Snowstorm: The crippling storm of December 22–23, 2004, brought record-breaking snowfall totals across southern Illinois, southern Indiana, and

Total Precipitation Percent of Mean
June 1, 1993 to August 31, 1993

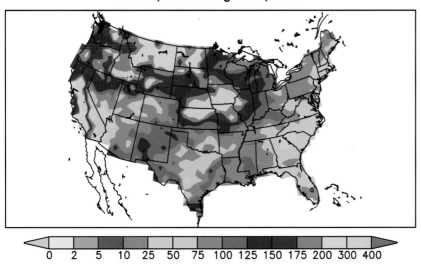

| 0 | 2 | 5 | 10 | 25 | 50 | 75 | 100 | 125 | 150 | 175 | 200 | 300 | 400 |

Midwestern Regional Climate Center
Illinois State Water Survey
Champaign, Illinois

FIGURE 7.9. Map of precipitation for summer 1993. Precipitation is expressed as percent of the 1971–2000 average. Blue shades indicate above average precipitation, while yellow and orange shades indicate below average precipitation. Figure courtesy of the Midwestern Regional Climate Center.

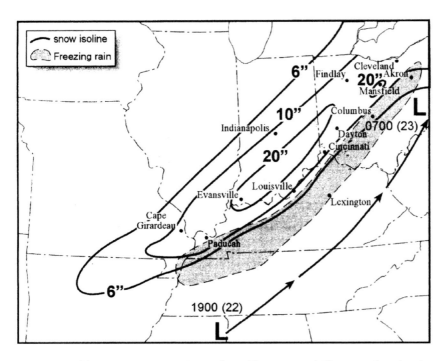

FIGURE 7.10. The 2004 snow storm. Figure from Changnon and Changnon (2005) and courtesy of the Midwestern Regional Climate Center.

western Ohio. A map of snowfall (figure 7.10) shows that a band of more than 20 inches occurred across southern Indiana, where amounts as high as 29 inches were recorded. Losses totaled nine hundred million dollars (Changnon and Changnon 2005).

A related area of research has focused on climate variability and trends. A key objective in these studies is to determine whether the climate of the Midwest is changing in any way. Some results of these studies include:

- There have been increases during the latter part of the twentieth century in the frequency of heavy precipitation events and associated losses from flooding.
- There is also evidence of high frequency heavy precipitation events early in the twentieth century, though not as high as in recent decades.
- The 1930s were characterized by the worst heat in the historical record.

The MRCC has been heavily involved in the development of new datasets. In the 1990s the MRCC led a project in collaboration with the region's state climatologists to digitize climate data taken during the period of 1895–1947. Prior to this project, these data were only available on paper or microfiche forms and they could not be included in computerized analyses. The MRCC is currently engaged in another project, in collaboration with the National Climatic Data Center, to do the same for data recorded during the 1800s.

INDIANA'S PRESENT CLIMATE—
THE STATE VIEW

The previous chapters contained many facets of Indiana's weather and climate often expressed in maps and diagrams. The purpose of this chapter is to bring together the many diverse characteristics of Hoosier climate and view it as a whole. Initially, the most important source of climate information in the state, the Indiana State Climate Office, is described. This is followed by an account of Indiana's climate written by a long-term climate specialist who has served the Indian State Climate Office in many capacities, including as its director. Few are as well informed about the climate of Indiana as Ken Scheeringa. Thereafter, various causes of local climates are examined and explained. The chapter ends with an examination of wind power in Indiana. The account provides many interesting and meaningful observations.

The Indiana State Climate Office

In 1954 the Weather Bureau Climatological Division established the National State Climatologist Program. The purpose of each state office was to document and study the climate of the state and respond to requests for information and data from individuals and groups. This federally funded program continued until 1973 when the program was terminated. However, each individual state was given the opportunity to take over the program and support it using state funds. It is fortunate for Hoosiers that the Department of Agronomy at Purdue University accepted the program as an extension function. It has continued to house the state climatologists as it has since 1956.

Over the years, the office has done a yeoman service with a very small staff. In recent years under the leadership of director Dr. Dev Niyogi and Associate Ken Scheeringa, its activities have expanded considerably. The website http://www.agry.purdue.edu/climate/contact.asp provides a considerable amount of information on Indiana climate, current climate maps, an entry for data requests, and useful links. Additionally, research

Table 8.1. Maximum and Minimum Temperatures in Indiana (°F)

Month	Max. Temp.	Year	Place
Jan	80	1943	Madison
Feb	83	1938	Shoals
Mar	91	1929	Madison
Apr	100	1930	Winona Lake
May	103	1934	Collegeville
Jun	111	1936	Seymour
Jul	116	1936	Collegeville
Aug	111	1936	Seymour
Sep	108	1953	Madison
Oct	100	1939	Madison
Nov	91	1933	Shoals
Dec	78	1982	Evansville

Month	Min. Temp.	Year	Place
Jan	-36	1994	New Whiteland
Feb	-35	1951	Greensburg
Mar	-19	1943	Goshen
Apr	1	1982	Goshen College
May	18	1978	Frankfort
Jun	30	1918	La Porte
Jul	37	1892	Marion
Aug	33	1946	Frankfort
Sep	21	1899	Prairie Creek
Oct	8	1906	La Porte
Nov	-10	1950	Wheatfield
Dec	-30	1924	Rochester

Source: Indiana State Climate Office website: http://www.agry.purdue.edu/climate/facts.asp.

projects are outlined including the preparation of an Indiana Atlas of Climate. Many of the state maps in this volume were from the working draft of the Atlas and are used with permission of the State Climate Office.

There is far too much data available from the office to include here, but Table 8.1 provides some interesting data, while figure 8.1 provides a map of the climate divisions of Indiana.

The Normal Climate

Ken Scheeringa

Indiana has an invigorating climate with strongly marked seasons. Winters are often cold, sometimes bitterly so. The transition from cold to hot weather can produce an active spring with thunderstorms and tornadoes. Oppressive humidity and high temperatures arrive in summer. Autumn is favored by many residents as a pleasant time of the year with lower humidity than the other seasons, and mostly sunny skies. Indiana's location within the

23. Climate Divisions

FIGURE 8.1. The climatic regions of Indiana have been used in reporting climate data for many years. Figure courtesy of the Indiana State Climate Office.

continent highly determines this cycle of climate. The Gulf of Mexico is a major player in Indiana's climate. Southerly winds from the Gulf region readily transport warm, moisture-laden air into the state. The warm, moist air collides with continental polar air brought southward by the jet stream from central and western Canada. A third air mass source found in Indiana originates in the Pacific Ocean. Due to the obstructions posed by the Rocky Mountains, however, this third source arrives less frequently in the state.

Temperature

Air temperatures in Indiana have a wide annual range due to the state's location and its natural characteristics. The state record observed daily maximum temperature is 116°F at Collegeville on July 14, 1936. The record minimum temperature is -36°F observed on January 19, 1994, at New Whiteland. January is typically the coldest month of the year with normal daily maximum temperatures ranging from 31 to 38°F north to south across Indiana. Normal January minimums range between 15 and 21°F north to south. July is the warmest month with daily maximums averaging 80 to 83°F and minimums 63 to 65°F north to south. Winters have been milder than usual during the last decade due to the strong influences of El Niño on Indiana weather. Prolonged severe hot and cold spells are uncommon in Indiana.

The dates of the last freezing temperature in spring and the first in autumn vary greatly from year to year. Two-thirds of the time they occur within a twenty- to twenty-four-day period centered at the mean date (see chapters 2 and 4). The average date of the last freezing temperature in spring ranges from the second week of April in extreme southwest Indiana to the second week of May in the extreme northeast. The trend of a later date toward the north is reversed in extreme northwestern Indiana, where the average date is about May 1 near Lake Michigan. In autumn the average date of the first temperature of 32°F or colder is from September 26 in the extreme northeast to October 26 along the Ohio River in the southwest.

Spring freezes end later and autumn freezes begin earlier in valleys and hollows along a given latitude. The gradual slope of the terrain upward from southwestern Indiana to northeastern Indiana results in lower minimum temperatures and shorter growing seasons in the east compared to the west at the same latitude. Muck soils in northern Indiana transfer soil heat poorly. These soils can freeze as early as late summer, resulting in a shortened growing season. Soil heaving due to frequent freeze-and-thaw cycles is most problematic in south central Indiana.

Precipitation

Average annual precipitation ranges from thirty-seven inches in northern Indiana to forty-seven inches in the south. May is the wettest month of the year with average rainfall between four and five inches across the state. Average rainfall decreases slightly as summer progresses. Autumn months are drier with three inches of rainfall typical in each month. Indiana winters are the driest time of the year with less than three inches of precipitation commonly received each month. February is the driest month of the year statewide, then precipitation increases in March and April as the spring soil moisture recharge season begins. On average, precipitation occurs every third day in Indiana.

Annual precipitation is adequate, but an uneven distribution in the summer occasionally limits crops. Mild droughts occasionally occur in the summer when evaporation is highest and dependence on rainfall is greatest for crops. Approximately one-third of the annual rainfall flows to the Mississippi or Great Lakes, mainly during cool weather. The soil usually becomes saturated with water several times during the winter and spring. Ground water storage is generally abundant in the north and central areas where glacial deposits cover ancient lake beds or streams. An underlying bed of limestone with shallow soils limits ground water storage in much of south central Indiana.

Floods occur in some part of the state nearly every year and have occurred in every month of the year. The months of greatest flood frequency are from December through April. The primary cause of floods is prolonged periods of heavy rains, although rain falling on snow and frozen ground is a contributing factor.

Average annual snowfall ranges from fourteen inches in southwest Indiana to seventy-six inches in the north central snowbelt near Lake Michigan. Snowfall amounts vary greatly from year to year depending on both temperature and the frequency of winter storms. Measurable snow typically begins in late November and ends by early April although the season can begin as early as mid-October and end as late as early May. In warm years snow may not begin until mid-December. At a given latitude in central and southern Indiana snowfall amounts increase toward the east because of the higher elevation.

Other Climatic Features

Cloudiness is least in autumn and greatest in winter. The sun is usually visible about 65 percent of daylight hours on summer days but only 30 percent of the time on winter days. The northern part of the state is cloudier than the south, particularly in the winter when the Great Lakes have their greatest effect upon the weather.

During daylight relative humidity is usually lower in the south than in the north. This is true for all seasons. However, the simultaneous occurrence of high temperatures and high relative humidity is most frequent in the south. This combination defines the heat index which is often in the uncomfortable zone during much of Indiana's summers.

Total evaporation from a water surface in a four-foot-diameter tank ranges from six inches at Valparaiso to eight inches at Evansville in July. Evaporation is three or four inches in the north in April and October, and a little higher in the south.

As discussed in chapter 2, severe storms which damage property and cause loss of life are most frequent in the spring, although tornadoes have occurred in every month of the year in Indiana. On June 2, 1990, thirty-seven tornadoes ripped through Indiana, the most on any one day in state history. Property damage is greatest from high winds during thunderstorms, while hail occasionally causes loss of crops over small areas during the summer.

Prevailing winds average near ten miles per hour and travel generally from the southwest during most of the year. Wind speeds in excess of one hundred miles per hour have been measured in Indiana near severe storms and tornadoes. During winter months winds prevail from a northerly direction and are more persistent. The land and sea breeze effect is prominent in the summer and the cooling sea breeze tends to reduce daytime maximum temperatures along the Lake Michigan shore and for up to a mile inland.

Recent Climate: The Common Sense Climate Index

Global warming in the past century amounts to only about 1°F (0.5°C). When a new global record temperature is set, it may exceed the previous record by only a few hundredths of a degree. What relevance, if any, do such small temperature changes have to most people?

A simple measure is needed of the degree to which practically noticeable climate change is occurring.

The Common Sense Climate Index is a composite index of climate quantities that are noticeable to the lay person. Positive values of the index refer to changes that would be associated with warming, and negative values with cooling. It employs climate indicators that are easily understood and defines a scale that will reveal when change should be noticeable above the level of natural climate variability.

The index is based on daily observations in the case of cities for which daily data are readily available. Other locations use only monthly mean data, which restricts the analysis to a small number of climate indicators. However, examination of results for cities with daily data shows that the index based on monthly data is usually similar to the composite index based on all climate indicators including daily data. The following list includes those indicators currently included in the index for cities with daily data. The index is based on the four seasonal mean temperatures for cities with only monthly data.

- Seasonal mean temperatures (four seasons)
- Degree days (heating season, cooling season)
- Frequency of extreme temperatures ("hot" summer days, "cold" winter days)
- Record daily temperatures (record highs, record lows)

Figure 8.2 shows the Common Sense Climate Index (upper graph) and actual temperatures (lower graph) for Indianapolis, 1880–2008. Periods of warming and cooling are clearly seen in the index graph, while using the actual temperature data, they are difficult to decipher.

Some Local Influences

The Urban Heat Island

How often have your heard a television weather forecaster saying "Temperatures in the city tonight will be in the mid-30s, but will fall to the upper-20s in outlying areas?" Or, "Afternoon temperatures will reach 90 degrees downtown and the mid-80s away from the city." These forecasts reflect the fact that the weather and climate of urban areas differ from the surrounding countryside.

Humans have a remarkable facility to alter the natural environment. Any change results in a modification of the ongoing natural processes at that site. Perhaps the most changed environment is the city; the constructed environment of a city creates a totally different climatic realm from that which occurred prior to its founding. Consider the following:

Concrete, asphalt, and glass replace natural vegetation

Structures of vertical extent replace a largely horizontal interface

Large amounts of energy are imported and combusted

Combustion of fossil fuels creates pollution

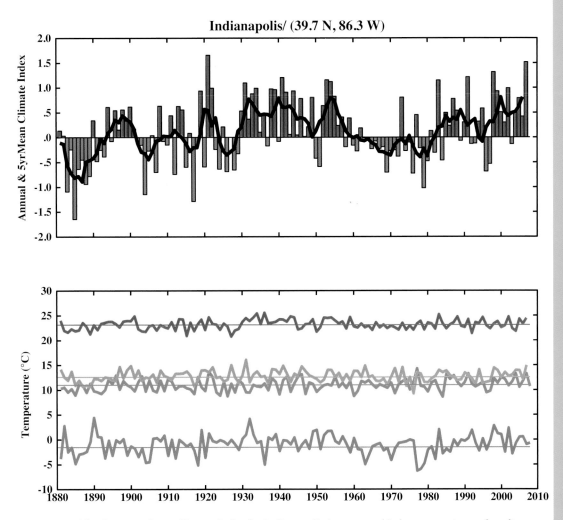

FIGURE 8.2. The Common Sense Climate Index for Indianapolis (upper graph) shows warming and cooling trends much more clearly than the average temperatures (lower graph) for the period from 1880 to 2008. In the upper graph, the vertical axis shows the index values; the red bars are annual values above normal; the blue bars are average annual values below normal; and the black line shows the five-year mean values. In the lower graph, the vertical axis shows the temperature (in °C); the blue line represents temperatures recorded in December, January, and February; the green line, those recorded in March, April, and May; the red line, those recorded in June, July, and August; and the orange line, those recorded in September, October, and November. Figure courtesy of the National Aeronautics and Space Administration (NASA).

These are related factors that modify the climatic process in the urban environment. Walls, roofs, and streets present a much more varied surface to solar radiation than undeveloped areas. Even when the sun is low in the sky, a time when little energy absorption occurs on flat land, vertical city buildings feel the full impact of the sun's rays. In the early morning and late evening, the city is absorbing more solar energy than surrounding rural areas (figure 8.3).

Thanks to a grant from the National Science Foundation, researchers from the Department of Geography at Indiana State University are looking into methods that could help large cities stay just a little bit cooler by planting trees to spread their shading branches over streets and sidewalks.

"Satellite images provide global coverage. By using remote-sensing measurements, we can do urban heat detection and monitoring for all major cities in the world. One of our goals is to apply the methodology we develop here in Indianapolis to other cities," said Qihao Weng, the research leader.

Indianapolis is planting plenty of trees these days. Officials are keeping records of each tree the city is planting and ensuring they are spaced in a way that is intended to provide the most benefit.

The researchers believe the layout of urban landscapes—from lawns, trees, parks, and cemeteries to streets, parking lots, and buildings—plays a role in determining the urban heat island effect. Indianapolis's White River State Park, for example, gently curves along the banks of its namesake. Weng has a theory that such a layout holds the key to reducing the urban heat island effect, even if by just a few degrees.

"Theoretically speaking, a park with a square or rectangular shape is going to generate more heat than a park of the same total area with a circular shape because it has more heat interactions with the surrounding area of impervious surfaces," Weng explained. Indianapolis is being used as a pilot project to develop methods for cities worldwide to use in addressing urban heat islands. One

Rain falling on urban and non-urban areas is disposed of quite differently. On building-free rural surfaces some of the water is retained as soil moisture. Plants draw upon this source for their needs and eventually return the moisture to the air. At the same time, standing water and soil moisture evaporate. Solar energy is required for this process. In the city, pavements and buildings prohibit entry of water into the soil and most of the rain water drains off quickly and passes to storm sewers with the result that water availability for evaporation is greatly diminished. Since solar energy is not used to transfer moisture to the atmosphere, it becomes available as a heating source for the city.

As areas of concentrated activities, cities are high consumers of energy, and enormous amounts of energy are imported to maintain their functions. This high use of energy means that much "waste" heat from factories, buildings, and transportation systems passes to the atmosphere to add to the warmth of the city.

The higher temperatures are best developed at night, when, under stable conditions, a heat island is formed. In passing from the center of the city toward the surrounding countryside, temperatures decrease slowly until, in rural areas, they remain about the same. The city is an island of warmth surrounded by cooler air.

Wind speeds in a city are, on average, lower than those in open surrounding areas because of the increased roughness of the urban fabric. A comparison of highest expected wind speeds in the city and its airport (usually located outside the city in flat, unobstructed land) shows this to be true.

Hills and Valleys

Many visitors to Indiana are often surprised to find that the state is not one large flat area! The well-known images of Indiana corn fields give no hint of the variation in elevation that occurs in the state. Indiana topography is characterized by flat plains in the northern two thirds of the state while in the south, hills, ridges, knolls, caves, and waterfalls abound. A few counties in far west central Indiana also exhibit the southern topography due to their location in the Wabash River bed. Land elevations range from 324 feet above sea level at the mouth of the Wabash River in the southwest corner of the state to 1,257 feet in far eastern central Indiana.

Most of the state is drained by the Wabash River system. The total drainage area of the Wabash is 33,000 square miles, of which 24,000 square miles are in Indiana. Other river basins are the Maumee in the extreme northeast, the St. Joseph (Lake Michigan) and Kankakee (Illinois River) in the north central and northwest, respectively, while some of the extreme south and southeast area drains into the Ohio River.

South central Indiana has the most rugged terrain and is home to the Hoosier National Forest. The Kankakee Valley in the extreme northwest slopes gently toward the west and drains what were formerly marshlands. Many small lakes abound in northeastern Indiana among numerous glacial moraines and hills. Tourism has become a

of the first steps is field studies that verify—or, in the case of the new subdivision, correct—information gathered from satellite images and remote sensing.

The research group is visiting and photographing each of 350 locations throughout Marion County and comparing temperature estimates generated by remote sensing with actual weather station readings on the ground.

"After we do this, we can tell the urban planning people, 'Hey, you can fix this problem by changing the shape, use or type of land cover'," said Hua Liu, a Ph.D. student in geography. "We cannot just have a huge parking lot with a huge building; we have to have some grass—to cool the temperature down."

The group believes Indianapolis and Indiana State University are leading the way in conducting such research and developing ways to minimize the effects of urban heat islands. Source: *ISU News*

FIGURE 8.3. Urban areas modify their local climates in many ways, ultimately leading to the urban heat island. Figure courtesy of John E. Oliver.

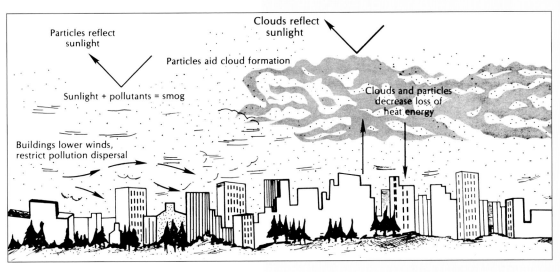

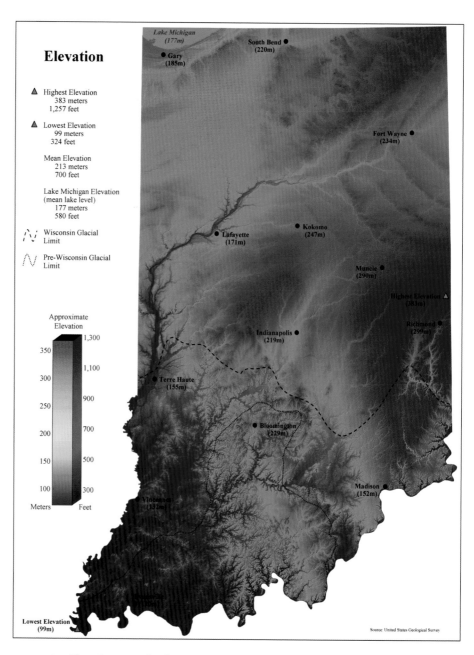

FIGURE 8.4. Elevation map of Indiana. Figure courtesy of the United States Geological Survey and the Geography Educators' Network of Indiana (GENI).

growing industry in southern Indiana while farming and manufacturing remain important on its northern and central plains.

Figure 8.4 shows elevations in the state, and the differences shown, while not large in comparison to western states, does have an influence on weather and climate.

A significant effect of elevation differences rests with the fact that temperature generally decreases with altitude. Given that the difference between the widely spaced highest and lowest points in the state is only 933 feet, the role of elevation is subtle and secondary to other factors. However, in local areas the variations in elevation do have an influence.

The usual decrease of temperature with height does not always occur, with the inversion, described in chapter 5, providing a significant example of how temperature close to the ground can be cooler than the air above. Differences in elevation can produce temperature differences over quite small areas, especially in hilly terrain. Consider a valley on a cold, clear, windless night. The ground will chill rapidly and lose its heat to the air immediately above. This heat will be conducted to the air above, and the layer next to the ground will become cooler. Eventually, the lowest layer of air will be at the same temperature as the ground, and colder than the air above. Cold air is denser than warm air, and the layer of cold air will flow down the sides of the valley to form a "pool" in the valley bottom.

If the dew point of the air in the valley pool is reached, then a dense valley fog might result. If the temperature of the air in the "pool" falls below 32°F, then frost will result in the valley bottom. Given such an effect, many orchard growers in hilly terrain locate their first layer of trees somewhat up the side of the valley to prevent frost damage.

The La Porte Anomaly

Zack Payne

For a short period of time the northern Indiana city of La Porte received the attention of both national television and print media. A study in the 1960s by Stanley Changnon of the University of Illinois showed that at that location, precipitation averaged 30 to 40 percent more than the surrounding area. Furthermore, the increase was attributed to the influence of the Chicago urban area. While the anomaly is no longer a topic of interest for the media, climatologists have long been interested in the phenomena as an example of the influence of urbanization in areas "downwind" of the possible source.

Figure 8.5 depicts the five-year moving average of precipitation that compares the occurrence of the La Porte Anomaly with the surrounding cities of Valparaiso and South Bend. The anomalous precipitation amounts at La Porte can clearly be seen, starting around 1930, increasing rapidly in 1940, and continuing through 1960. The La Porte area was also shown to have more counts of heavy precipitation than any of the surrounding stations.

FIGURE 8.5. The five-year moving averages of annual precipitation at La Porte and two other stations, 1910–2003. The 1940–1960 anomaly at La Porte is clearly seen but is shown to have diminished in recent years. Figure courtesy of Zack Payne.

For the reanalysis of the five-year moving average plot, the annual rainfall for the three stations was collected for 1966–2003. These data were plotted with the original data for comparison. As seen in figure 8.5 the La Porte five-year moving precipitation average spikes higher than the other stations for a few years in the 1990s, but the South Bend line also does the same. This more recent La Porte line is not nearly as high as the anomaly spike, and it does not last as long as the original anomaly.

Comparing reanalyzed data to the original, the anomaly seems to have moved even westward toward Valparaiso from 1974–1984. The next twenty years do not show any more signs of the anomaly, indicating that the anomaly may have moved out of the region, or the precipitation rate just might have faded back to the average.

The overall results of this study indicated that the La Porte Anomaly does not exist anymore in the northern Indiana area. Newly created maps of precipitation in the La Porte area do not show precipitation levels significantly higher than any of the surrounding areas. Perhaps the abnormal rainfall patterns for the earlier period from 1940–1960 is best explained by Chicago area urbanization and the many smoke-haze days of the period.

Wind Power of Indiana

Steve Stadler

Wind is a tangible aspect of Indiana's climate. Cooling summer breezes, damaging thunderstorm winds, and winds piling snow in drifts are well known to Hoosiers. In the late 1800s and early 1900s, small windmills were used to pump water in agricultural applications. Before widespread rural electrification, some farms had devices such as winchargers which produced a couple of hundred watts of power with battery backup. The connection of all of Indiana to power grids and the increasing use of electricity in businesses and residences ended reliance on wind power. As electricity costs rise and technology improves, wind is being seriously examined as a future power source for Indiana. It is not anticipated that wind power will ever supply a majority of Indiana's energy, but it could well provide a few percent of the energy supply as it does in other states.

Wind power is a function of wind speed and air density. It is measured as wind power density in watts per square meter, independent of any wind machine size or model. Trees and buildings represent mechanical friction and slow the wind so that wind power density is greater with distance away from earth's surface. Wind speed is the main determinant of wind power density and is tied to wind power density through a cubic function; that is, as wind speed increases a small amount, wind power density increases by a large amount. If wind speed doubles, for example, wind power increases eightfold (that is, $2^3=2\times2\times2=8$). The density of the air is a secondary factor controlling wind power density, and is dependent on altitude and temperature. In Indiana, altitudinal variations are not great enough to significantly affect the density of air. Lower temperatures make the wind more powerful at any particular wind speed because in any given volume of air there are more air molecules. Indiana's temperature-caused air density differences might result in a 1–2 percent wind power density change from winter to summer.

Figure 8.6 is a schematic of a typical wind turbine. The modern utility-scale turbine is one of the largest machines ever built. The diameter swept by the blades approximates the wingspan of a Boeing 747 jet (more than two hundred feet) and the nacelle (housing for the working parts) commonly sits one hundred and fifty feet or more above ground level. Blades are made of fiberglass-reinforced polyester, start rotating at wind speeds of

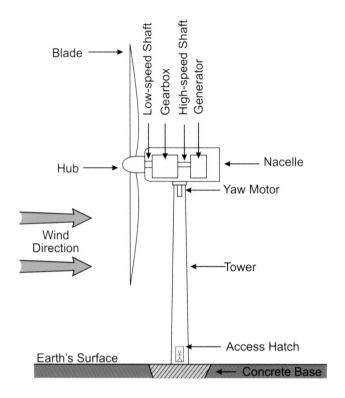

Blade

Low-speed Shaft
Gearbox
High-speed Shaft
Generator

Hub

Nacelle

Yaw Motor

Wind
Direction

Tower

Access Hatch

Earth's Surface

Concrete Base

FIGURE 8.6. (*above*)
Schematic of a utility-scale
turbine (not to scale). Such a
turbine generates more than
a million and a half watts.
Figure courtesy of Steve
Stadler.

FIGURE 8.7. (*left*)
A ten thousand watt
home wind generator.
Photo courtesy of Bergey
Windpower.

approximately seven miles per hour, and complete ten to twenty revolutions per minute. Turbines are always pointed into the wind and have continuous pitch control so that the blades can slice the air at the currently optimal angle. Inserted into the hub, the blades turn a shaft connected to a gearbox which, in turn, is connected to a high speed shaft driving a generator. Each of these wind turbines can generate over a million and a half watts, enough to power over two hundred homes. Wind turbines used in home and farm applications have smaller blades and shorter towers (figure 8.7), and generate a few thousand watts. Wind economics are such that the larger the turbine, the lower the cost per watt generated; however, some farm and home wind applications are inexpensive enough to clearly make economic sense.

The judicious placement of turbines is accomplished using a number of considerations. One is a detailed knowledge of the local wind power density and this requires wind measurements at proposed sites. Higher ground is favored over lower ground because of generally higher wind speeds on higher ground. Also, land surface sloping away from a turbine provides an extra measure of lift to propel the blades as the wind comes toward a turbine. Forested landscapes are not preferable because trees slow wind near the earth's surface. Domestic and wild animals have been shown to tolerate the presence of large turbines. Birds and bats are of concern because of the possibility of collisions with the blades, although bird collisions tend to be relatively infrequent with the current utility-scale turbines. Viability of utility-scale wind development is dependent upon distance to transmission lines and the loads already carried by those lines. Finally, public perception comes into play. In some locations turbines have been welcomed while in others they have been banned by law. In Indiana it is likely that wind development will be looked upon favorably as a way to bring income and electricity to rural areas. The economics of farm and home turbines are quite variable by location within Indiana, but there are many sites feasible for their placement.

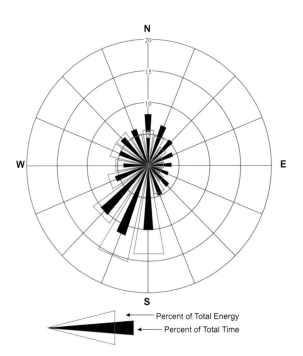

FIGURE 8.8. Annual wind direction and power roses for 295 feet at Goodland, Indiana. The blue color is the percentage of time the wind comes from each of the sixteen wind directions. The red color is the percentage of total wind power from each of the sixteen wind directions. Figure courtesy of the Indiana Office of Energy and Defense Development.

Figure 8.8 is an example of a wind rose. Wind roses are used to convey long-term wind information important to the placement of wind turbines. This rose represents annual wind on a ninety meter (295 foot) tower near Goodland in Newton County. The concentric circles each increment 5 percent of the total time wind blows from a particular direction. Winds are measured in sixteen compass directions with winds blowing from the north at the top of the figure. The blue wind rose indicates the percentages of the year winds blow from each of the directions and is representative of much of Indiana in that the dominant

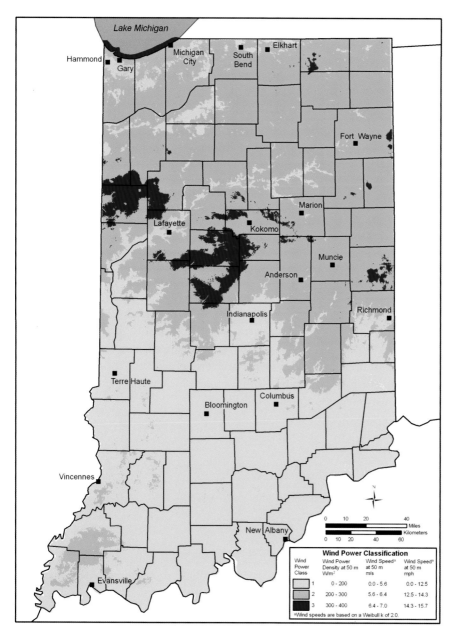

FIGURE 8.9. Wind power density map of Indiana. Figure courtesy of the National Renewable Energy Laboratory.

wind directions are from the southwest and west southwest. For instance, wind blows from each of the southwest and west southwest directions about 10 percent of the time. In terms of potential wind power, the red wind rose indicates that the pattern is more pronounced with about 34 percent of wind power derived from the combination of the southwest and west southwest directions. The difference between wind direction occurrence and wind power direction occurrence is because long-term wind speeds are not equal by direction. Modern turbines automatically adjust to wind speed and direction through the turning of the hub into the wind, yet the wind rose is important to wind power development because terrain and vegetation are not uniform in all directions. Indiana wind roses suggest particular attention be given to the fetch of air in the southwest and west southwest directions. That is, ideal turbine locations might feature lack of trees and have the steepest slopes faced in those directions.

The wind industry uses wind power density to assess the wind power potential of a location. As wind power installation has rapidly developed around the United States, state wind maps have been created through the use of long-term wind data meshed with terrain and vegetation patterns. Figure 8.9 represents the average annual wind power potential of Indiana at fifty meters (161 feet) above the surface; although fifty meters is below the height of typical turbine hubs, this figure accurately represents the spatial variability of wind power potential across Indiana. In general, northern Indiana has greater wind power than southern Indiana. There are two reasons for this. First, northern Indiana is more often under passing weather disturbances; and, second, northern Indiana has less land covered by friction-causing trees.

Nationwide, class 4 wind power (wind power density at 50 m of 400–500 W/m²) is usually thought to be adequate for utility-scale wind generation. Indiana does not have any large areas of class 4 winds. Indiana has most of its class 3 wind in Benton, Clinton, Boone, and Randolph Counties, with smaller areas in adjacent counties and in the northeast. Of potential importance is a small strip of class 3 and class 4 wind offshore in Lake Michigan. In the near future, anticipated improvement in efficiency means that utility-scale wind generation might be viable over the class 3 areas of Indiana. Of note is the fact that most of Indiana can currently be tapped for home and farm generation. The scale of figure 8.9 does not show localized areas of hilltops with few trees which might be entirely satisfactory for small wind installations in southern Indiana.

INDIANA'S PAST CLIMATES

Using the instruments of today, minor changes or trends in weather are routinely monitored and analyzed. However, the period for which weather instruments have been available is but a tiny fraction of time in earth's history. To understand current climates and to predict future climates, it is absolutely essential that they be considered in the framework of climatic change over geologic time. To achieve this, the climates of the past must be reconstructed. This process reflects much painstaking, detective-like research because much of the evidence is based upon the relationship of climate to other environmental processes and past life. Prior to presenting an account of how climate has varied, this chapter deals first with the way in which climates are inferred.

Reading the Past

The reconstruction of climates over time is a fascinating puzzle. Instrumental records have become available only in very recent times and information about earlier climates require the use of proxy data, observations of other variables that serve as a substitute or proxy for the actual climatic record. Proxies are ancient archives of climate. In recent years, the development of highly sophisticated methods of analyzing and dating materials has led to great advances in reconstructing the past.

What, then, are some methods by which proxy data are derived? As shown in Table 9.1, they are considerable. Some, like ice cores are found far from present-day Indiana, but they still play a role in interpreting the past. Ice cores from Greenland, for example, are analyzed for different forms of oxygen isotopes—for example, oxygen-18 (^{18}O) and oxygen-16 (^{16}O)—that occur in different ratios in ice formed at different temperatures. This permits the actual temperature that existed more than 100,000 years ago to be obtained and provide a general picture of the prevailing northern hemisphere climates of that time. More locally, the evidence of the Great Ice Age that ended some 10,000 years ago is plentiful in Indiana.

Table 9.1. Outline of Proxy Methods for Reconstructing Past Climates

Proxy method	Definition	Spatial extent	Timescale	Benefits	Some limitations
Ice cores (^{18}O, gases, dust)	Cored ice from ice caps and glaciers	Greenland, Antarctica, some high-altitude and high-latitude glaciers	Several 100,000 years	• Long records over multiple millennia • Wide range of information possible • Annual/seasonal analysis possible	• Small area of earth's surface • Cold climates only • Detailed dating can be imprecise • Accuracy reduced with depth
Dendrochronology	Width and density of tree rings	Poleward of 30°N or S where cold season stops growth	Up to about 10,000 years	• Details about temperature and moisture • Correlates well with measured data • Potential for ^{18}O and ^{14}C analysis • High-quality data record	• Limited to high elevations or mid-to-high latitudes • Limited to trees with annual growth rings • Dependent on growing season months
Coral	Growth rings and chemistry from tropical coral massives	Tropics with shallow seas	Several hundreds of millennia possible	• Tropical compliment to tree rings • Can provide precise dating • Continuous sampling possible • Provide teleconnection details (ENSO)	• Tropics only • Limited spatial distribution • Long records are rare • Seasonal cycle dominates
Pollen	Pollen species from undisturbed lake and coastal cores	Mainly mid-to-high latitudes where trees and grass grow	Normally a few millennia	• Widespread use • Long records possible • Can indicate both temperature and moisture	• Dating can be uncertain • Pollen identification can be difficult • Depends on how vegetation responds to climate

Proxy method	Definition	Spatial extent	Timescale	Benefits	Some limitations
Speleothems	Stalactites and stalagmites in cave environments	Cave environments; depends on water flow in sedimentary rock	Normally a few millennia	• Wide range of proxies possible (chemistry; trace elements) • Can indicate changes in water cycle and atmospheric circulation	• Limited locations • Dating can be uncertain • Difficult to interpret climate variable
Varved sediments	Sedimentation of marine and lake organic remains	Areas with high sedimentation rates and strong seasonal changes	Normally a few millennia	• Annual and seasonal resolution possible • ^{18}O analysis • Faunal assemblies	• Best in closed-basin glacial lakes • Limited distribution
Historical records	Written records, diaries, phenology, crop harvests, etc.	Can be wide ranging, but usually mid-to-high latitudes	Up to about 1,500 years	• Wealth of information possible • Wide range of potential sources • Can provide highly detailed information	• Patchy in space and time • Can be anecdotal and inaccurate • Requires cautious interpretation • Often emphasis on extremes

Source: Bridgman and Oliver (2006). Note: ^{18}O, also known as oxygen-18, is a stable and naturally occurring isotope of oxygen used to date oxygen-containing samples. ^{14}C, also known as carbon-14 or radiocarbon, is a radioactive isotope of carbon used to date carbon-containing samples.

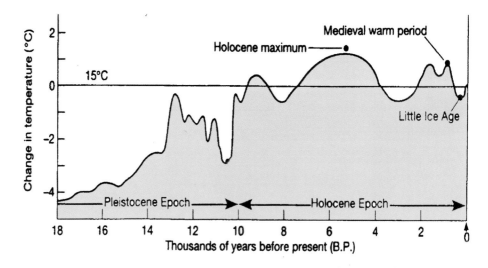

FIGURE 9.1. Since the end of the Ice Age (Pleistocene), global temperatures have fluctuated. The warmest times were 4000 to 6000 years before present (Holocene maximum) and during Medieval times. Curves are exaggerated to show changes.

Given all the evidence available, it is possible to reconstruct the general climate of mid-latitudes in the northern hemisphere. Figure 9.1 shows a generalized reconstruction for the period since ice retreated from North America. Significant changes are seen in temperatures.

In looking at past climates of Indiana, source information rests in large part upon the research of the Indiana Geological Survey located at Indiana University; as part of their publications, the accounts provided by Anthony Fleming are interesting and understandable. This may be seen on the website at the address given in the list of references for chapter 9. Susan Berta of Indiana State University significantly contributed to the following account.

Ice Age Indiana

The rich agricultural lands and predominance of lakes and wetlands found throughout portions of Indiana have been left to us as evidence of "recent" climatic change brought about by Indiana's Ice Age. Evidence of climate change can easily be viewed by examining the landscape of Indiana and using a little imagination. The most recent dramatic changes to the landscape in Indiana caused by ice took place during a time when the global climate appears to have been 7°F (4°C) to 18°F (10°C) cooler worldwide and daily temperatures at our latitude fluctuated seasonally in a similar manner as they do today. That time period is called the Ice Age. An ice age refers to a time when periods of much colder climates dominated and widespread glaciation occurred on earth.

Glaciation

Glaciation occurs whenever snow and ice accumulated in a thick enough layer or blanket to begin to move under the influence of gravity. Snow and ice, left to accumulate on the landscape, will undergo periodic melting and refreezing forming an ice material referred to as firn. As the landscape continues to be covered in an increasingly thicker blanket of snow, firn, and ice, the mass begins to move under the influence of gravity; hence, a glacier is created. Ice flowing across a landscape can scour and deposit material in its path. Vast flows of ice are called ice sheets, a type of glacier, and they are known to have covered exposed land surfaces for millions of years at a time. Specific reference to the Ice Age refers to widespread glaciation during the Pleistocene Epoch of the geologic time scale. The Pleistocene began 2 million years ago and ended approximately 10,000 before present (BP). Ice sheets covered large portions of North America and Eurasia, and all of Greenland and Iceland. The ice sheet that spread across the North American continent during this time is called the Laurentide Ice Sheet and it reached its maximum coverage approximately 20,000 years ago.

Before the glaciers occupied Indiana, the northern reaches were likely similar to the rugged landscape found in south central Indiana today. The major drainage ways that now exist in Indiana (Wabash River, Ohio River, and Lake Michigan) are a result of the glacial ice eroding, moving, and depositing rock and ice materials. Prior to the Ice Age, the major drainage way for northern Indiana was the Teays River, an ancient stream later destroyed by the glaciers. Crossing central Indiana, its valley was estimated to have been 200–400 feet deep. The Great Lakes that we know of today did not exist and an ancient river system drained toward the St. Lawrence.

Ice Age Events in Indiana

During the Pleistocene, glaciers pulverized the high points and filled in the low points much as a bulldozer would while reworking the same piece of ground over and over again. Scientists have identified numerous lobes of ice that moved, rarely in sync with one another, across the North American continent leaving sags and/or glacier-carved basins where ice accumulations were once very thick. Four large basins that once contained major lobes of ice that impacted Indiana's landscape include three lake basins (Lakes Michigan, Huron, and Erie) and a bay (Saginaw Bay). These lobes of ice transported shattered and pulverized rock and ice material generally referred to as till. Tills (or till units) vary primarily by differences in texture and color and, therefore, can be identified for each ice lobe. Scientists use the till units to aid in distinguishing ice lobe activity across Indiana.

Ice sheets are known to grow in size (advance) and melt so as to reduce in size (retreat). During the Ice Age there were four major episodes when the Laurentide Ice Sheet is known to have made major advances in North America. Those episodes are referred to as the Nebraskan, Kansan, Illinoian, and Wisconsinan Advances. Each major advance was followed by a warming period that caused the ice sheet to retreat as far north as Canada. The oldest of these advances is the Nebraskan and we have yet to find a locale where Nebraskan deposits are known to exist in Indiana. Approximately 700,000 years ago, during the Kansan glacial period, the Ice Sheet occupied over two-thirds of the state of Indiana while the glacial advance during the Illinoian (at least 500,000–140,000 BP) exceeded that of the Kansan advance (figure 9.2). Evidence on the landscape of these events may be found in the soil profiles and glacial till identified as unsorted rock material carried by the glaciers. There are two prime locations for observing these older glacial tills: near Cagle's Mill in southern Putnam County and at Bean Blossom Reservoir in northern Monroe County. For instance, at the northeast end of the emergency spillway at Cagle's Mill, glacial till from the Wisconsinan time period occupies, on average, the top three feet of earth followed by twenty-three feet of Illinoian till and then thirty-two feet of Kansan till for a total thickness of till equal to approximately fifty-eight feet of sediments laid down during the Pleistocene.

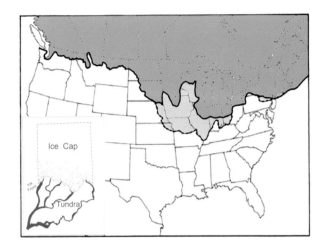

FIGURE 9.2. The continental ice sheets in North America. The dark blue shading shows the most recent advance (the Wisconsinan advance) and the light blue shows evidence of earlier glacial advances. The inset shows the ice front in Indiana and the harsh climate (tundra) to the south.

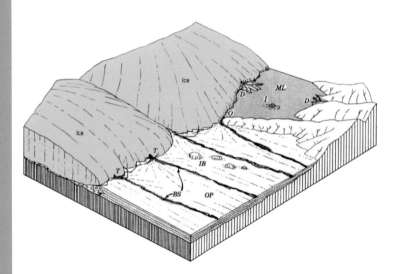

FIGURE 9.3A. A schematic view of the landscape at the edge of an ice sheet such as that which covered northern Indiana (upper figure). Figure adapted from Strahler (1951).

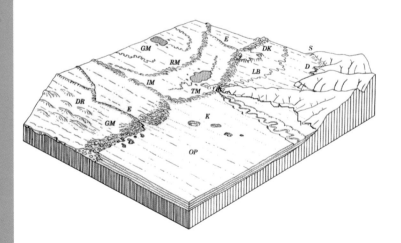

FIGURE 9.3B. The resulting landscape features after the ice has withdrawn (lower figure). Figure adapted from Strahler (1951). The explanation of the features is provided in Table 9.2.

As noted, much of Indiana's scenery is a result of deposits by advancing and retreating ice. Of note are moraines, hills composed of glacial till, that mark the locations where glaciers either "paused" on the landscape or where stagnating or "stillstands" of ice caused moraines to form. Glaciers act much like conveyor belts; even though the glacier itself may not be moving across the landscape, the conveyor belt continues to carry and deposit glacial material at the end of the ice lobe. Rock and ice simply pile up to great depths whenever a glacier stalls for a time. Figure 9.3 illustrates a landscape covered in ice (upper) and the resulting scenery after the ice has retreated (lower). The terminal moraine separates the relatively flat outwash plain from the hilly glacial deposits that is sometimes known as "knob and kettle." There are many other features of ice advance and retreat in Indiana, and, while highly interesting, it is not appropriate to describe them all here. However, to provide a general guide to some of the features figure 9.3 shows some examples. The key to the abbreviations in the figure is given in Table 9.2. Many books and articles provide details of the features shown and some are listed in the references for this chapter.

Conditions in what we now call Indiana during the glacial advances would have been hard to imagine. Anthony Fleming, in his description of the Ice Age in Indiana, writes:

> Environmental conditions near this ice sheet would have been harsh indeed! Possibly as much as a half-mile to a mile thick, the imposing ice front would have been the scene of frigid cascades of meltwater and icy lakes amidst a landscape of thick, cold mud put in motion by constant freeze-thaw activity and water. The sense of desolation would have been

Table 9.2. Explanation of Features Illustrated in Figure 9.3

Upper Figure—Ice in Place	
T	Meltwater flowing from tunnel beneath ice
BS	A braided stream—a stream without a clear single channel that frequently changes position
OP	Outwash plain consists of debris-deposited braided streams
IB	Ice blocks surrounded by glacial debris
ML	A marginal lake formed by damming of meltwater
I	An iceberg in the glacial lake
D	A delta formed where streams flow into lake
O	The marginal lake outlet
Lower Figure—After Ice Melt	
IM, RM, TM, GM	Forms of glacial moraines: the interlobate (IM), recessional (RM), terminal (TM), and ground (GM). The terminal moraine is often the most visible feature.
E	A long sinuous mound formed beneath the ice; an esker.
DR	Drumlins, rounded hills that occur in groups or swarms
D and DK	Remnants of the delta formed in the marginal lake
S	A shoreline formed by the marginal lake
LB	The flat lake bottom contains layers of deposited material
OP	The outwash plain
KL	Kettle lakes formed when ice blocks were surrounded by debris before the ice melted

How time is recorded in the past is a complex issue that changes depending upon how we measure time for the different proxy records. Our longest records of lake sediment and ice cores are dated by comparison to paleomagnetic reversals and radiometric techniques, such as radioactive element decay measured in thorium-230/uranium-234 and carbon-14 (radiocarbon). Each of these dating methods is imperfect. Today, we also realize that radiocarbon progressively diverges from calendar dates the further back in time one measures. These dates can be reported as radiocarbon years before present (radiocarbon yr BP), which means uncalibrated years that can be as much as 2,000 years off once the sediment is 10,000 years old. To avoid this error, radiocarbon dates have been calibrated with dendrochronology (that is, dating using tree rings) for the past 10,000 years providing good estimates (+/-50 years) around the real calendar date of the event. When lake core sedimentary records have calibrated dates they are reported as calibrated years before present (cal. yr BP), which means that they have been corrected by comparison with dendrochronology. When these records refer to "before present" or "BP" they actually assign the present as a set point in time at AD 1950. Because atomic weapons testing in the modern era created an unusually high amount of radiocarbon (^{14}C) in the atmosphere, radiocarbon dates are not accurate for recent material. Due to this phenomenon, the fact that radiocarbon dating was invented by Willard Libby in 1949, and the need to have a standard point in time from which to measure, radiocarbon dates are counted from AD 1950; for example, 50 cal. yr BP actually corresponds to AD 1900.

compounded by the constant pounding of katabatic winds—cold blasts of air that roar off of the glacier as air masses passing over the ice surface are abruptly chilled and sink rapidly. In fact, an ice sheet of this scale would have probably made its own local weather. Sudden cooling of warm, moist masses of Gulf air encountering the cold ice sheet would have created torrential summer rains and near-constant "glacier-effect" snow in winter. During midsummer, at its southern extent in central Indiana, it is likely that one could have stood next to the glacier wearing shorts and a tee-shirt in weather similar to what we would experience today on a warm day in May! But the incessant cold wind pouring off the ice would be a constant reminder of the bitter winter to come. (Fleming 1997)

The Last 1,000 Years of Indiana's Climate

James H. Speer

This account focuses on the last 1,000 years where scientists can study past climate changes with some detail, even to the point of reconstructing annual temperature and precipitation throughout this period. During this time scale, two main events have been recorded known as the Medieval Warm Period (also known as the Little Climatic Optimum), AD 1100–AD 1300, and the Little Ice Age, AD 1600–AD 1850 (figure 9.4). These events were first recorded in Europe, and scientists are still debating about the amount of effect observed in North America. On an annual basis, drought events that last a decade have been recorded (such as the Dust Bowl event) as well as short-term wet periods. All of this climatic variability helps researchers determine the natural variability of climate and with that information, make better predictions about future climate change. The final conclusion from all of this data is that climate varies at all spatial and temporal scales, making prediction in this system difficult, but our most accurate re-cords are beginning to show an increase in temperature and precipitation in Indiana as predicted by climate change models that take anthropogenic (that is, human-caused) warming into consideration.

As already noted, paleoclimatology research helps put modern climate and its trends into a long-term perspective of how the system behaves and changes through time. The use of multiple proxy records of cli-mate help researchers examine different lines of evidence that provide checks on all of the climate reconstructions and strengthen the validity of interpretations from these records. These multiple proxies also provide different spatial and temporal resolutions, but together they provide a fairly clear picture of past climatic fluctuations which can be used to understand modern climate variation.

Trees in temperate climate zones put on one ring per year which is why dendrochronology is used to calibrate the radiocarbon record. Trees are organic, carbon-containing matter subject to radiocarbon dating and they are an independent record of time with one year recorded by each ring. So when dendrochronological dates are re-ported, they are reported as calendar year dates. Historical records also are based on the Christian calendar, so are reported just as years.

FIGURE 9.4. Climate over the last 1,000 years in the northern hemisphere. The Little Ice Age followed a period of warmth called the Little Climatic Optimum, now more commonly known as the Medieval Warm Period. Since the end of the nineteenth century, global temperatures have been warming. Curves are exaggerated to show changes. Figure courtesy of James Speer.

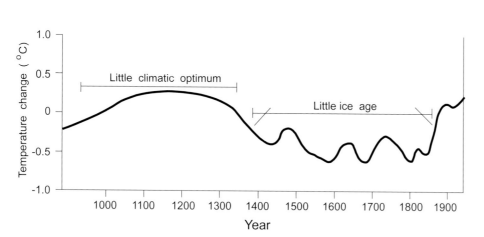

Indiana: Past Climate Interpretations

In understanding Indiana's climate, it is useful and necessary to draw on studies from surrounding states. Climate trends can be understood on a regional scale because weather patterns often cross several states, areas at the same latitude and elevation often have similar climatic conditions, and weather patterns do not stop at arbitrary state lines. Some of the best information on Indiana's climate in the past 1,000 years, therefore, will come from research conducted from Michigan to Kentucky, and information as far away as North Carolina can provide some insight into major events that have happened in the past as well.

Paleoeclimatic records indicate that there have been several climatic fluctuations in the past 1,000 years in Indiana in which conditions were different from today. For example, evidence indicates that Indiana was affected by global climatic conditions including the Medieval Warm Period and the Little Ice Age. Such records are important as a long-term comparison with current climate trend.

A Pollen Study

In 1974, Alice Williams used pollen data and sediment changes from Pretty Lake in northeastern Indiana, north of Fort Wayne, to examine climate change during the past 15,000 years. Her lake core data demonstrate a general warming trend since the beginning of the Holocene at 10,500 radiocarbon yr BP until about 1,685 radiocarbon yr BP. This warm and dry period resulted in the dominance of an Oak-Hickory forest throughout northern Indiana. At 1,685 radiocarbon yr BP, Williams documents a transition to a cool, moist climate resulting in a Beech-Maple forest in this region. At approximately 670 radiocarbon yr BP a warmer, drier climate caused a return to an Oak-Hickory forest (Table 9.3). At 150 radiocarbon yr BP, an obvious increase in the pollen of weeds such as ragweed, plantain, and dock was the result of land clearing by European settlers for agricultural purposes. A farming signature with dominance of these plants often obscures the climatic signal from many modern pollen records. This local data set shows the global warming trend in the Holocene after the Last Glacial Maximum that is similar to other records around the world. The warm and dry period peaked at about 2,500 radiocarbon yr BP, after which time Williams documented a subsequent cooling followed by another warming period. The modern record is obscured by the farming signature and the low temporal resolution of this study. Other research at a finer resolution provides further information about more recent climatic fluctuations.

A Lake Core Study

One study completed by Booth and Jackson in 2003 used high resolution data for recent climate change by examining the fossils of testate amoebae from lake cores over the past 3,500 cal. yr BP. Testate amoebae are small aquatic organisms that develop an interior shell of calcium carbonate or silica which can be preserved for long periods of time, providing information about water conditions during their lifetime. Quantification of different species of testate amoebae and knowledge of their preferred habitats enabled the researchers to document fluctuations between drier and wetter conditions in Minden pond in Michigan throughout the lake core.

The amoebae document a wet period from 1600 to 1100 yr BP, transitioning to dry conditions starting at cal. 950 yr BP (AD 1000), with a marked increase in moisture in the

Table 9.3. Interpretation of Pretty Lake Core Data

Years before Present (BP)	Climate	Vegetation
10,500 to 1685	A warming trend eventually leading to warm, dry conditions	Oak-Hickory forests
1685 to 670 (Approx. AD 265 to AD 1280)	Cool, moist conditions prevail	Beech-Maple forest
670 to 150 (Approx AD 1280 to AD 1800)	Return to warm, dry conditions	Oak-Hickory forest
AD 1800–present	Warm, dry conditions	Oak-Hickory forest with evidence of forest clearing

last 50 cal. yr BP. This evidence is supported with pollen data also from the lake core showing a general decline of beech and sphagnum moss over the last 950 cal. yr BP indicating drier conditions overall. The influx of Europeans into America and the increase in farming is evident over the last 150 cal. yr BP in this lake, as is an increase in ragweed, grass, goosefoot/lambs quarter, and plantain pollen, similar to the Williams (1974) chronology.

1,000 Years—Concluding Comments

As these varied records of climate for the Midwest show, Indiana experienced the Medieval Warm Period as well as the Little Ice Age which are global climatic events. Modern climate temperature seems to be increasing relative to the beginning of our meteorological records from the late 1800s, but this could be a recovery from the Little Ice Age. Most of the global warming is predicted to occur in northern latitudes and higher elevations where they are currently being observed. In Indiana the predicted climate changes are not likely to be as extreme, but we do see that Indiana responds to global climatic trends and seems to be recording some slight warming and increase in precipitation as predicted by general circulation models.

Indiana Diaries and Letters

Cameron Craig

In reconstructing climates of the past one hundred and fifty years, the writings of some Hoosiers provide quite a lot of information. The study of these writings is both informative and interesting. In 1870 the United States Weather Bureau established a standardized system of weather observations where observers used the same calibrated thermometers and rain gauges, and recorded the weather at particular times. This method made it easier to analyze and interpret current weather conditions across the nation for the purpose of forecasting. Prior to the standardization, some people used various instruments that were not similar in construction or observed the conditions infrequently. Today, climatologists can obtain and use instrumental data from the National Climate Data Center as far back as 1895. However, investigating the climate prior to this year requires the use of a substitute, a proxy. Diaries and personal journals left by Indiana citizens long ago can provide climatologists insight into the weather conditions, flood frequency, and the dates of frost and blossoming of trees.

A great many of people in the nineteenth century wrote about their daily adventures in diaries. They described who they met, what they bought and their costs, chores they completed, the number of livestock and bushels of grain they sold, and the weather to complete the entry. In regards to the weather descriptions, some people simply wrote the conditions in relative terms. Others provided actual numeric data if they had the instruments to do so. Most diaries had relative and comparative information—that is, they contain terms like "hottest," "warmest," or "coldest"—because it was sometimes expensive to purchase a thermometer even if one was available.

Interpreting the Records

Historical documents in the form of personal diaries, journals, lists, and other unpublished writings are most often held in archival institutions such as the Indiana Historical Society's William H. Smith Memorial Library in Indianapolis, Indiana. Since most of the collections housed in this archive are fragile and historically significant, documents are non-circulating and review of the documents must be done at the library. In some cases, the reviewer must wear gloves to protect the documents from the oil from fingers. The interpretation of handwritten documents can be tedious due to the ornate calligraphy used in the period. However, the opportunity to look back at a time unfamiliar to the climatologist is enlightening and rewarding.

Using historical documents requires the climatologist to become a climate detective in order to reconstruct a prior climate. The use of one person's diary is not enough to gain scientific knowledge of a former climate; therefore, several diaries within the same region are required. In order to determine the accuracy of one statement recorded by a person, another must be used that describes the same situation or event. For example, one person's diary entry states that "it was the worst flood in history." The climatologist asks him or herself: in relation to what? Did this person live throughout history? Probably not; therefore, the use of another witness's entry must be used. "It rained for three days and the river engulfed the bridge into our town, which has never happened in my lifetime," another witness observed. The climatologist cannot take for granted that every description provided by people will be the most accurate story. The use of several diaries can reduce overstatements, understatements, and confirm what actually occurred. Knowing the background information about the diarist is also important to the climatologist. A person who was born and raised in Florida for twenty years who then moved to Indiana would probably exaggerate the snowfall they experienced. Not having an idea what a blizzard was, if their new region experienced snow flurries, they would probably have written, "It was a blizzard."

Digging into each diary entry, the words used by former inhabitants can be tricky. When a person uses the word "hot" today it does not necessarily have the same meaning to another person. This is also the case between centuries. Someone who uses the word "hot" to describe the temperature on August 8, 1841, does not necessarily have the same meaning in our current century. "Hot" back then could have meant 86°F. Today, it could mean 102°F. There is one article in the *Terre Haute Daily Express* written on July 14, 1858, regarding the temperature on July 2 and 4, 1776, titled "Hot Days":

> 2nd of July 1776, the day the Declaration of Independence passed Congress, the thermometer in Philadelphia stood at 77° at 9 hours and 40 minutes, and at 74° at 9PM. On the 4th of July, when it was signed, the thermometer stood at 68° at 6AM, 72° at 9AM, 76° at 1PM, and at 73½° at 9PM. These were "times that tried" men's bodies as well as "souls." Yes, the blood that coursed through the veins of the signers of the Declaration of Independence stood at 76° all the time the sons of England went down, and never have been at the freezing point yet.

This statement was made eighty-two years after the signing of the Declaration of Independence. The problem with this statement is that we do not know who regarded it as hot. Did someone state that it was hot? If so, who? In this context, this writer probably thought that it was hot, at least in the context of his own time.

Example of an Indiana Climate Reconstruction

Two observers from the mid-nineteenth century are examined to demonstrate how a climate reconstruction can be accomplished for Indiana.

Mr. Elisha King was a dedicated weather observer and his data are extremely valuable—he provided consistent weather observations every day for nine years. He was a man of self-determination. Mr. King operated a farm handed down from his father and took a "lively interest in Sabbath-school work." One could easily be drawn into his daily life through reading his diary. His weather observations for the first few years were extremely detailed. The word descriptors he used throughout the diary numbered twenty-two.

Throughout his diary, Mr. King described the overall weather most frequently as being normal or rather "pleasant" with "cold" being second. What can be deduced from this is that the relative temperature on his farm for the period between 1854 and 1862 was cool. Mr. King described, on eleven occasions, that the temperature was the "coldest." Some of these occasions were written as "the coldest yet this season." On one occasion, July 17, 1855, Mr. King indicates the temperature being "the hottest." It is unfortunate that Mr. King did not have a thermometer in order to provide us a glimpse into what he considered the hottest. However, Mr. Samuel Shirk of Brookville, Indiana, did provide the actual temperatures in his daily diary and indicated July 17, 1855, as being 98°F. This particular instance is remarkable. For two people, unknown to each other and located one county apart, to provide similar observations is a triumph in historical reconstruction.

Samuel Shirk served as a major in the War of 1812 and was county commissioner of Franklin County at one time. Mr. Shirk's diary measures only four inches wide and four inches high. He had two annual diaries in which he recorded the temperature, business transactions, and brief daily activities.

Furthering the correlation between specific observations recorded by Mr. King and Mr. Shirk, Table 9.4 indicates several dates where the two have similar observations.

It is interesting that on July 15, 1855, Mr. King indicated that the temperature was "very hot and sultry" and Mr. Shirk provided the actual temperature as 98°F. However, referring to the table below, Mr. King described the temperature as being "the hottest" for a lower temperature on July 17. There are two possible explanations for this: first, there is

Table 9.4. Examples of Indiana Historical Data

Date	Elisha King	Samuel Shirk
1/25/1855	Snowing AM/Quit Snowing/ Warmer	Snowing/20°F/2 inches deep—the most this winter
2/3/1855	The coldest day yet. Clear nearly all day.	Clear—0°F Coldest day.
2/27/1855	Cold and Clear	Clear thermometer a shade above zero
4/17/1855	Warm—the warmest	Hazy 86°F
7/15/1855	Very hot and sultry	Clear 98°F
7/17/1855	The hottest and clear	Clear 94°F

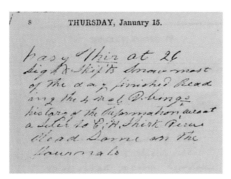

FIGURE 9.5. Examples of diary entries. Figure courtesy of Cameron Craig.

Example of Mr. Samuel Shirk's writing. [Transcription] "Thursday, January 15. Hazy Thir [Thermometer] at 26. Light skifs of snow most of the day finished reading the 4 vols D[illegible]s history of the Reformation, wroat a leter to E. H. Shirk [illegible] Read some [illegible] the Journals." Samuel Shirk Collection, Indiana Historical Society.

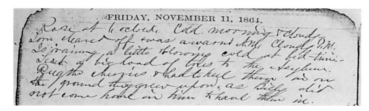

Example of Mr. George Beeler's writing. [Transcription] "Friday, November 11, 1864. Rose at 6 o'clock. Cold morning & cloudy. Soon cleared off & was a warm[er] A.M. Cloudy P.M. Is raining a little blowing snow at bed-time. Sent a big load of [illegible] to the Asylum. [Illegible] cherries & had to h[a]ul them in [illegible] the ground they grew upon as Billy did not come home in time to haul them in." George Beeler Collection, Indiana Historical Society.

no way that a person can tell the difference between the change of a degree of temperature without using instruments; and, second, it is possible that other factors played a part in Mr. King's perception (unknown to him) of the change in humidity and/or wind speed. This is a problem when dealing with relative temperatures in historical documents. The actual temperature may not change, but the conditions felt by the human body and perceived in the mind can lead one to use different descriptions.

Based on Mr. King's observations, it is interesting to note an overwhelming trend toward colder temperatures. In addition, his observations can provide insights into how long or short a season was by looking at the negative values that indicate cold temperatures. In this particular example, most winters were more notable than the summers except for the winter of 1857–1858. The 1857–1858 winter season was a shorter season when compared to the other winters. This would imply that the winter of 1857–1858 was warmer than typical for the period covered by Mr. King.

Although historians are traditionally attached to historical documents for their reconstruction of particular events in the past, climatologists and other physical scientists can also use documents to learn about the natural world and its evolution over time. Beyond the descriptions of weather and numerical data, diaries provide us with an experience through the eyes and hands of those before us—a vital component in understanding the future state of our cultural and natural world.

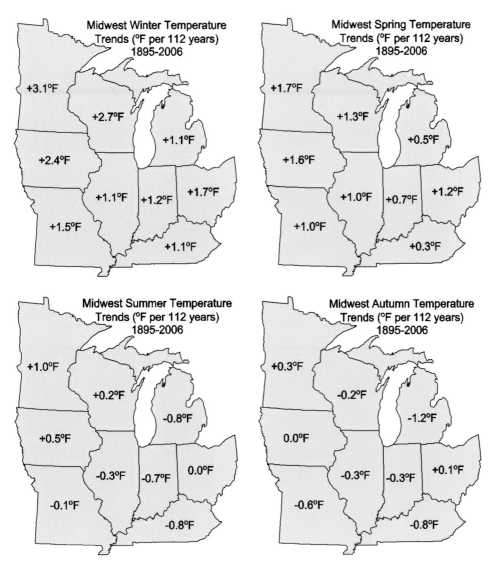

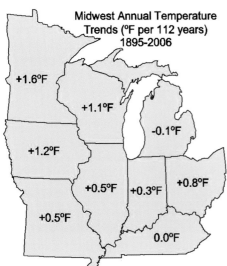

FIGURE 9.6A. Maps of the Midwest average annual and seasonal temperature trends between 1895 and 2006. Figure courtesy of the Midwestern Regional Climate Center.

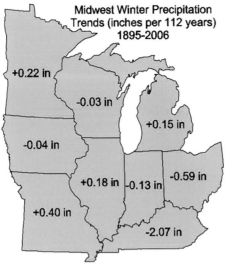

Midwest Winter Precipitation Trends (inches per 112 years) 1895-2006

+0.22 in
-0.03 in
+0.15 in
-0.04 in
+0.18 in
-0.13 in
-0.59 in
+0.40 in
-2.07 in

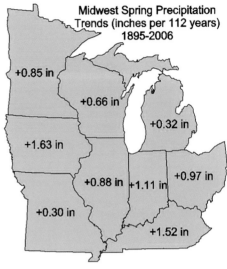

Midwest Spring Precipitation Trends (inches per 112 years) 1895-2006

+0.85 in
+0.66 in
+0.32 in
+1.63 in
+0.88 in
+1.11 in
+0.97 in
+0.30 in
+1.52 in

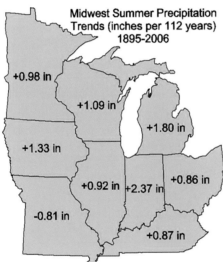

Midwest Summer Precipitation Trends (inches per 112 years) 1895-2006

+0.98 in
+1.09 in
+1.80 in
+1.33 in
+0.92 in
+2.37 in
+0.86 in
-0.81 in
+0.87 in

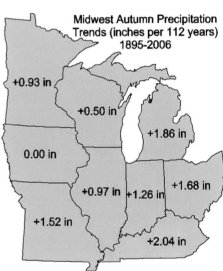

Midwest Autumn Precipitation Trends (inches per 112 years) 1895-2006

+0.93 in
+0.50 in
+1.86 in
0.00 in
+0.97 in
+1.26 in
+1.68 in
+1.52 in
+2.04 in

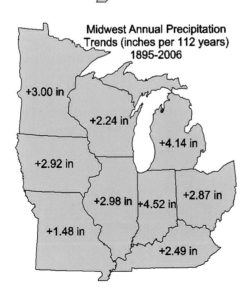

Midwest Annual Precipitation Trends (inches per 112 years) 1895-2006

+3.00 in
+2.24 in
+4.14 in
+2.92 in
+2.98 in
+4.52 in
+2.87 in
+1.48 in
+2.49 in

FIGURE 9.6B. Maps of the Midwest average annual and seasonal precipitation trends between 1895 and 2006. Figure courtesy of the Midwestern Regional Climate Center.

The Last 100 Years

A brief look at climates over the last one hundred years was given in chapter 8. To look at the region a little more closely, the Midwestern Regional Climate Center has produced a series of maps showing how temperatures and precipitation have changed for the period of instrumentation, 1895 to 2006.

The analysis shows that the climate of the Midwest has changed over time since the beginning of modern records in 1895. Maps of the state average annual and seasonal temperature and precipitation trends between 1895 and 2006 are shown in figures 9.6a and 9.6b. Temperature trends are given in degrees Fahrenheit change and precipitation trends are reported as inches of precipitation change, both for a period of 112 years. The state averages used in calculating the trends came from the National Climatic Data Center Climate Division Dataset.

The temperature maps show that both seasonal increases and decreases occur. Spring and winter temperatures have increased while those of summer and fall have decreased. The overall result is a decrease in average annual temperature of 0.3°F.

The mapped precipitation values indicate that the state has experience increased rainfall for 112 years. The annual trend shown, 4.52 inches, is an appreciable amount with only winter showing a decreasing trend. These recent changes in climate are discussed in chapter 10 to provide context for possible future climate states.

10

INDIANA'S FUTURE CLIMATES

To obtain a daily weather forecast requires an understanding of atmospheric processes, accurate measurements, extensive computer knowledge, and great skill. Forecasters provide us with twenty-four-hour forecasts that are mostly quite reliable. The forecasts for forty-eight-hour and seventy-two-hour future weather conditions, while still very good, are somewhat less reliable. As one looks further into the future, the incredibly complex atmosphere can do quite unexpected things, and the longer ahead the forecast, the less certain it becomes.

Consider now the problems encountered when attempting to make a climatic forecast. Certainly, there is some success in forecasting the number of storms that will occur in the hurricane season and in the published thirty-day and sixty-day outlooks, but how about the very long-term future climate? Forecasting what the global climate or the climate of Indiana will be like in twenty-five, fifty, or one hundred years is a formidable task, but it is one that is necessary to undertake. The potential effects of global warming, as discussed later in this chapter, are so wide ranging that it becomes a necessity to attempt to forecast future climates.

The IPCC and Protocols

Before discussing the potential impacts of climate change upon Indiana, it is necessary to view the context in which climate change is considered on a global scale. Of major importance in dealing with the science of climate are the Intergovernmental Panel on Climate Change (IPCC) and the implemented protocols that address the political and economic ramifications of climate change.

It is of particular significance that the IPCC was awarded the 2007 Nobel Peace Prize. The work of the IPCC is guided by the mandate given to it by its parent organizations, the World Meteorological Organization and the United Nations Environment Programme. IPCC principles contain specific procedures for the preparation, review, acceptance, approval, adoption, and publication of IPCC materials. The three main classes of IPCC materials are IPCC reports, which include assessment, synthesis, special, and methodology reports; technical papers; and supporting materials, such as proceedings of IPCC workshops.

The IPCC has three working groups. Working Group I assesses scientific aspects of climate systems and climate change; Working Group II studies the vulnerability of human and natural systems to climate change and options for adapting to them; Working Group III assesses options for limiting greenhouse gas emissions and deals with economic issues. A task force also studies national greenhouse gas inventories. Much more detail and information is available on the IPCC website (http://www.ipcc.ch/).

Protocols dealing with climate change have an interesting history. At a meeting in Rio de Janeiro in 1992 the United Nations Framework Convention on Climate Change (UNFCCC) was adopted. Its purpose was to combat global warming by stabilizing the emissions of greenhouse gases. The Kyoto Protocol or, more properly, the Kyoto Protocol to the United Nations Framework Convention on Climate Change, is a proposed amendment that was adopted in 1997 at a Conference of Parties (COP) in Kyoto, Japan.

The Kyoto Protocol commits developed countries to reduce emissions of carbon dioxide, methane, and nitrous oxides and to phase out hydrofluorocarbons, sulfur hexafluoride, and perfluorocarbons over a thirty-year period. It reaffirms the idea that developed countries must supply technology to other countries in climate-related studies and related projects.

The protocol was endorsed by 160 countries and will become binding provided that 55 countries, including developed nations responsible for most global emissions, ratify the accord. By 2002 some 104 countries had signed the agreement. Signing the agreement is essentially symbolic, and for the protocol to become fully effective the UNFCCC requires that parties complete their "instruments of ratification, acceptance, approval or accession."

Since the Kyoto meeting, other COPs have attempted to resolve problems and issues, not always with success. Over time, and with appropriate political activity, full ratification may occur.

Modeling the Future

The only way to predict future climate is to develop computer models. The first step is to create a model that reflects the current global weather and climate system and then to modify that model to show the impact of changes that may occur within the system. Clearly, the sophistication of the models precludes a full description here, but it is useful to know the way in which they are constructed and used.

To derive a global model, the planet is divided into a three-dimensional grid. Equations, based upon the laws of physics, fluid motion, and chemistry, are used to replicate the existing atmospheric conditions at each grid point. This becomes exceedingly complex, for at each point air motion, radiation, heat transfer, moisture content, and surface characteristics are needed. The interactions between these are also considered. Clearly, a supercomputer is the key to solving equations and the multitude of interactions. The most recent models include the interactions between the atmosphere and oceans, ice covered surfaces, and land (figure 10.1).

Once a satisfactory representation of the current climate is obtained, the computations are stepped forward for a pre-selected time. Additionally, some of the values may

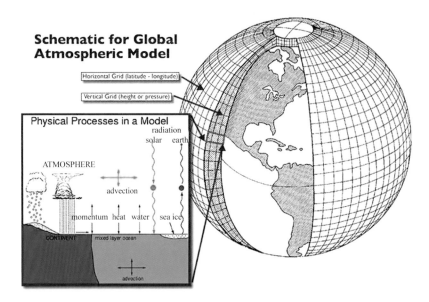

FIGURE 10.1. To derive a model for future climates, earth is divided into a three-dimensional grid. For each cell within the grid it is necessary to reproduce the ongoing physical processes. Figure courtesy of the National Oceanic and Atmospheric Administration (NOAA) and the National Aeronautics and Space Administration (NASA).

be altered; for example, one can double the present-day carbon dioxide levels in order to see how this affects the atmosphere. This is the way in which the impact of humans upon future climates may be assessed.

The accuracy of climate models is limited by the grid resolution used to represent the surface of the globe. Also, of course, is the ability to describe in mathematical terms the complicated processes that occur. However, despite some imperfections, models can simulate the current climate and its variability with some accuracy. Significant model improvements with more accurate representations and more capable supercomputers will certainly provide highly reliable scenes of future climates.

Global Warming

Of the many environmental problems of today, global warming probably heads the list. There is considerable evidence to substantiate the fact that the earth's climate is warming, and many scientists and scientific organizations have concurred that the earth is warming at an unusually fast rate.

Evidence of global climate change is present in many different forms. The most obvious and mathematically supportable evidence is in the temperature record. Analysis of temperature data gathered over the planet indicates rapidly rising temperatures over the past one hundred and forty years. Table 10.1 provides some data for the twentieth century which illustrates the temperature rise, particularly toward the end of the record. For example, the twentieth century was the warmest century in the last millennium, and temperatures in the world's oceans are also increasing, not simply at the surface, but at considerable depths. There are many temperature dependent phenomena which also indicate that the earth is warming. They include the following:

- Earth's mountain glaciers are melting
- Antarctica's ice sheets are breaking up
- Sea level is rising
- The temperature of the global ocean is rising
- Northern hemisphere permafrost is melting
- Arctic pack ice is thinning and retreating
- The tree line in mountain ranges is moving upward
- Many tropical diseases are spreading toward the poles and to higher elevations in the tropics

There are many books, articles, and websites devoted to global warming with perhaps the best-known and most reliable information coming from the IPCC.

Given that global warming is occurring, a question that may occur to the even the most casual reader is "why is this happening?" It was shown in the last chapter that the climate of the earth has changed appreciably over geologic time, and that as little as ten

Table 10.1. Twentieth Century Global Warming Observations

Global Temperature Data

- 1998 was the warmest year on record
- Seven of the ten warmest years on record have occurred since 1990
- The 1990s were the warmest decade on record
- The 1980s was the second warmest decade on record
- The ten warmest years on record have occurred since 1983
- The mean temperature of earth has increased about 1°F (0.5°C) in the twentieth century
- The twentieth century was the warmest century of the millennium

Ice Melting Observations

- The edge of the West Antarctic Ice Sheet is shrinking at the rate of 400 feet each year
- The Larsen Ice Shelf on Antarctica disintegrated in January 1995
- Much of Antarctica's Larsen B and Wilkes ice shelves disintegrated in 1998–1999
- The average elevation of glaciers in the Southern Alps of New Zealand moved upward 300 feet in the twentieth century
- In the Tien Shan Mountains of China, glacial ice shrank nearly 25 percent in the past 40 years
- In the Caucasus Mountains of Russia, half of all glacial ice melted away in the past 40 years
- In the Garhwal Himalayas of India, glaciers are rapidly retreating
- In the Andes Mountains of Peru, glacial retreat increased seven fold from 1978 to 1995
- In the Bering Sea, the area of sea ice shrunk 5 percent in past 40 years
- In the Arctic Ocean, the area of sea ice shrunk by 14,000 square miles since 1978
- The largest glacier on Mt. Kenya almost completely melted away in the twentieth century
- The Bering Glacier in Alaska is retreating
- The glaciers in Glacier National Park, Montana, are melting rapidly
- Glaciers in the Alps Mountains of Europe shrank by about 50 percent in the twentieth century. Gruben and Aletch glaciers are among those most rapidly melting

Source: Oliver and Hidore (2002).

To examine future climates of regions of the earth requires the use of models that provide a more detailed view than global models; to obtain that higher resolution requires a huge amount of computer time and space. There are a number of models that look at future climates of the United States, but one of special interest has been generated here in Indiana.

The Purdue Climate Change Research Center improved the resolution of future images of the United States by analyzing areas half as large as those used earlier; their model used areas 25 km to a side as opposed to the 50 km of previous models. This allowed them to discern climate regions of the United States in more detail.

The model assumed a doubling of greenhouse gases and the performance of the model was checked using data from 1961 to 1985. The following observations may be drawn from their results.

1) The area of Illinois and north of Kentucky, essentially the northeastern United States, will experience summers that will be much longer and hotter than they currently are.

2) The south, essentially the Gulf Coast area, will experience a changing set of conditions, with long dry spells alternating with heavy rainfall events.

3) The deserts of the Southwest will receive even less summer rainfall than now. Additionally, extremely hot conditions will increase significantly and make human habitation and occupation in the desert area even more problematic.

thousand years ago the northern parts of the state were ice covered. So is this current change part of the natural cycle that has occurred over the millennia? The answer is that it could be, but given the rapidity with which the warming is occurring coupled with observations about the atmosphere's contents, there appears to be an additional cause: an enhancement of the Greenhouse Effect.

It may be recalled that in chapter 1 it was pointed out that the warmth of the earth's atmosphere is due to the earth's heat being trapped by greenhouse gases. The theory of an enhanced Greenhouse Effect is that if more greenhouse gases are added to the atmosphere, more heat will be trapped and the earth will get warmer. One important greenhouse gas is carbon dioxide, a by-product of burning fuels. If the level of carbon dioxide in the atmosphere is carefully monitored (as it has been in Hawaii for more than half a century), then changes can be observed. Figure 10.2 shows the increase of carbon dioxide concentrations in the atmosphere over time. The ups and downs of each year represent the seasonal changes in photosynthesis. What stands out from the chart, however, is the increase in carbon dioxide content over time. This increase, together with additions of other greenhouse gases to the atmosphere, may account for the current rapidity of warming earth's atmosphere.

The models of the atmosphere are used to assess just how much warming may occur with additions of specified amounts of greenhouse gases. Most models agree that the warming may have significant worldwide and regional consequences. As the following *Of Interest* note below suggests, significant changes could occur in the United States with a doubling of the current carbon dioxide levels.

Given that global climate change is occurring, our concern at this point is how such changes may influence Indiana and the Midwest.

Indiana's Future Climate

There are a number of interesting sources that discuss potential climate change in Indiana. While most agree that change will occur, slight differences in the amount of change reflect the use of different models. Given the difficulties of producing model future climates for smaller regional areas, Indiana's future climate is often considered in the context of the future climate of the Midwest. The data and accounts published in 2000 by the United States Global Change Research Program are used here to provide illustrative examples of climate projections; their results are based upon two models, one from the Canadian Centre for Climate Modelling and Analysis, and one from the Hadley Centre for Climate Prediction and Research in the United Kingdom.

Over the twentieth century the northern portion of the Midwest, including the upper Great Lakes, has warmed by almost 4°F (2°C), while the southern portion along the Ohio River valley has cooled by about 1°F (0.5°C). Annual precipitation has increased, with many of the changes being quite substantial, as much as 10 to 20 percent increases over the twentieth century. Much of the precipitation has resulted from an increased rise in the number of days with heavy and very heavy precipitation events. There have been moderate to very large increases in the number of days with excessive moisture in the eastern portion of the Midwest.

During the twenty-first century models project that temperatures will increase throughout the Midwest and at a greater rate than has been observed in the twentieth century. Even over the northern portion of the region, where warming has been the largest, an accelerated warming trend is projected for the twenty-first century with temperatures increasing by 5 to 10°F (3 to 6°C). The average minimum temperature is likely to increase as much as 1 to 2°F (0.5 to 1°C) more than the maximum temperature.

4) The continental United States as a whole will see a greatly diminished length of winter and overall warmer temperatures.

As computers become faster and permit gigantic sets of data to be handled and analyzed, models such as this will improve and provide even greater details about the potential changes resulting from global warming.

Over the next century, climate in Indiana may change even more. For example, based on projections made by the IPCC and results from the United Kingdom's Hadley Centre's climate model (HadCM2), a model that accounts for both greenhouse gases and aerosols, by the year 2100 temperatures in Indiana could increase by 2°F in summer (with a range of 1–4°F), 3°F in winter and spring (with a range of 1–6°F), and 4°F in fall (with a range of 2–7°F). Precipitation is estimated to increase by 10 percent in winter and spring (with a range of 5–20%), 20 percent in fall (with a range of 10–40%), and 30 percent in summer (with a range of 10–50%). Other climate models may show different results, especially regarding estimated changes in precipitation. The frequency of extreme hot days in summer is expected to increase along with the general warming trend. It is not clear how the severity of storms might be affected by these changes, although an increase in the frequency and intensity of summer thunderstorms is possible.

Precipitation is likely to continue its upward trend at a slightly accelerated rate; 10 to 30 percent increases are projected across much of the region. Despite the increases in precipitation, increases in temperature and other meteorological factors are likely to lead to a substantial increase in evaporation causing a soil moisture deficit, reduction in lake and river levels, and more drought-like conditions in much of the region. In addition, increases in the proportion of precipitation coming from heavy and extreme precipitation are very likely. Figure 10.3 provides a graphic image of the changes using the two models.

The Environmental Protection Agency (EPA) has used the IPCC findings released in their report of 2001 to outline the future climate of Indiana. Their findings are very much in accord with those described above and, for comparison purposes, are given in the *Of Interest* at left.

Impacts of Climate Change in Indiana

There are many studies estimating the impacts of future climate change. That provided by the Union of Concerned Scientists has examined future climates in some detail and has provided a summary of the impacts of the changes in Indiana. Some of their findings are outlined here.

Agriculture

Given the importance of agriculture in Indiana's economy it is imperative that some estimate of the future be obtained. As noted earlier, impacts will concern not only higher temperatures but also changes in both precipitation cycles and the incidence of severe weather.

Increased temperatures, together with increased carbon dioxide, could well produce higher crop yields; these will be enhanced by a longer growing season. However, some of benefits may be nullified by hotter and drier

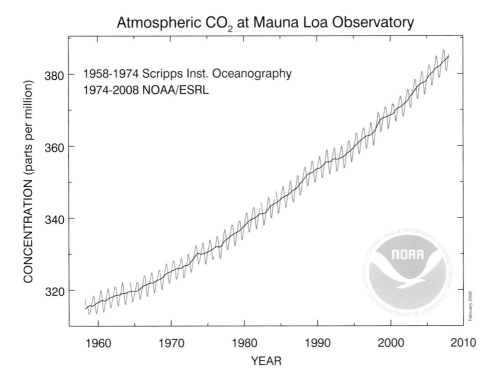

FIGURE 10.2. Carbon dioxide in the atmosphere has increased over time. Figure courtesy of the National Oceanic and Atmospheric Administration (NOAA).

summers, potentially longer droughts, and severe weather during the growing season. Another negative outcome related to the higher temperatures is the potential for greater pest infestation. In their jointly produced report titled *Confronting Climate Change in the Great Lakes Region,* the Union of Concerned Scientists and the Ecological Society of America note several consequences of higher temperatures:

> "First, warmer and shorter winters will allow more southerly pests such as corn earworms and fall armyworms to expand their range northward. Indeed, such a shift already appears to be happening with bean leaf beetles, which not only feed on soybeans but also serve as vectors for a virus that causes disease in soybeans. Second, warming will increase the rate of insect development and the number of generations that can be completed each year, contributing to a build-up of pest populations . . . Increased pests may also drive farmers to use more pesticides or related chemicals, placing an additional burden on water quality." (Kling et al. 2003)

This burden may also include additional soil erosion and runoff as a consequence of the increased rainfall.

Water Supply

Indiana depends heavily upon Lake Michigan and groundwater for its water supplies. Even without any change in climate, the growing population of the state will add additional pressure to water needs, and competing demands will be challenging. Climate change

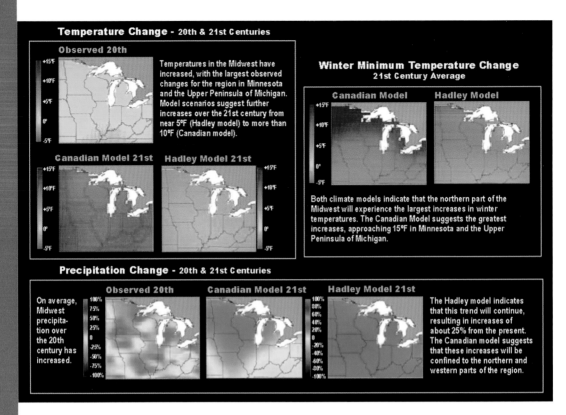

FIGURE 10.3. Changes in temperature and precipitation using observed data for the twentieth century and changes indicated by two models for the twenty-first century. Figure courtesy of the United States Global Change Research Program.

will complicate the situation for runoff patterns, and groundwater recharge will reflect the changing temperature, evaporation, and heavy rainfall patterns.

It is anticipated that there will be a decrease in lake and reservoir levels. Increasing temperatures will lead to less ice cover in winter resulting in more evaporation. Associated with falling lake levels and greater runoff from increased summer storms is a reduction in groundwater recharge. The reduction in lake and groundwater levels, together with population increase, will require careful political and economic planning to meet future water demands.

Human Health

Human health concerns related to climate change result from a complex set of interacting human and environmental factors. These concerns are particularly serious for the elderly and other vulnerable populations (the very young, the poor, and those whose health is already compromised). Climate change models project that, during the current century, extreme heat periods and severe storms are likely to become more common in the Great Lakes region. As a result, air and water quality, extreme heat periods, and storm-related risks could increase for residents of Indiana.

It is projected that the number of days with temperatures in excess of 90°F will increase considerably. The ill effects of high temperatures were considered earlier in the discussion of heat waves, and such effects will no doubt increase with climate change. On the positive side, however, is the expected decrease in health risks due to cold weather.

The EPA assessments point to higher air pollution levels in the years ahead. Of consequence is a potential increase in harmful ground-level ozone. The complex chemistry that leads to ozone formation occurs more rapidly at higher temperatures.

Changes in rainfall patterns and intensity can create major problems in stormwater capacities, and increase the prospect for waterborne infectious diseases. Again it is instructive to turn to the account of the Union of Concerned Scientists and the Ecological Society of America for an explanation of this.

"Extreme rainstorms can swamp municipalities' sewage and stormwater capacities, increasing the risks of water pollution and waterborne infectious diseases . . . For example, some waterborne infectious diseases such as cryptosporidiosis or giardiasis may become more frequent or widespread if extreme rainstorms occur more often. One of the best known examples of a *cryptosporidium* outbreak occurred in Milwaukee in 1993 after an extended period of rainfall and runoff overwhelmed the city's drinking water purification system and caused 403,000 cases of intestinal illness and 54 deaths. Milwaukee's drinking water originates in Lake Michigan." (Kling et al. 2003)

The occurrence of many infectious diseases is strongly seasonal, suggesting that climate plays a role in influencing transmission. St. Louis encephalitis outbreaks in the Great Lakes region, for example, have been associated with extended periods of temperatures above 85°F (29°C) and little rainfall. Some vector-borne diseases, such as Lyme disease or, more recently, West Nile encephalitis, have expanded widely across the region. While this spread is attributed largely to changes in the use of land, future changes in rainfall or temperatures could encourage greater reproduction or survival of the disease-carrying insects, which include ticks and mosquitoes.

In summary, the science of studying global climate change has advanced considerably in recent years and different climate models are now providing a similar outlook for future climates. There is little doubt that we are now in a period of global warming; while the role of humans in this is contested by some, it certainly seems that the addition of greenhouse gases by human activity is playing a role in the change.

A changing climate has significant ramifications for the future of Indiana. One hopes that both political and economic strategies can be implemented in order to meet the challenges of a modified environment.

APPENDIX

Conversion Tables

Temperature Conversion

To convert from Celsius to Fahrenheit read left to right
To convert from Fahrenheit to Celsius read right to left

°C	°F	°C	°F	°C	°F	°C	°F
50	122.0	27	80.6	4	39.2	-19	-2.2
49	120.2	26	78.8	3	37.4	-20	-4.0
48	118.4	25	77.0	2	35.6	-21	-5.8
47	116.6	24	75.2	1	33.8	-22	-7.6
46	114.8	23	73.4	0	32.0	-23	-9.4
45	113.0	22	71.6	-1	30.2	-24	-11.2
44	111.2	21	69.8	-2	28.4	-25	-13.0
43	109.4	20	68.0	-3	26.6	-26	-14.8
42	107.6	19	66.2	-4	24.8	-27	-16.6
41	105.8	18	64.4	-5	23.0	-28	-18.4
40	104.0	17	62.6	-6	21.2	-29	-20.2
39	102.2	16	60.8	-7	19.4	-30	-22.0
38	100.4	15	59.0	-8	17.6	-31	-23.8
37	98.6	14	57.2	-9	15.8	-32	-25.6
36	96.8	13	55.4	-10	14.0	-33	-27.4
35	95.0	12	53.6	-11	12.2	-34	-29.2
34	93.2	11	51.8	-12	10.4	-35	-31.0
33	91.4	10	50.0	-13	8.6	-36	-32.8
32	89.6	9	48.2	-14	6.8	-37	-34.6
31	87.8	8	46.4	-15	5.0	-38	-36.4
30	86.0	7	44.6	-16	3.2	-39	-38.2
29	84.2	6	42.8	-17	1.4	-40	-40.0
28	82.4	5	41.0	-18	-0.4	—	—

Length Conversion

Miles (Mi) and Kilometers (Km)

Mi	=	Km	Km	=	Mi
1	=	1.61	1	=	0.62
2	=	3.22	2	=	1.24
3	=	4.83	3	=	1.86
4	=	6.44	4	=	2.46
5	=	8.05	5	=	3.11
6	=	9.66	6	=	3.37
7	=	11.27	7	=	4.35
8	=	12.88	8	=	4.97
9	=	14.48	9	=	5.59
10	=	16.09	10	=	6.21
50	=	80.47	50	=	31.07
100	=	160.9	100	=	62.14

GLOSSARY

Acid precipitation Precipitation which is more acidic than natural precipitation. Natural precipitation has an average pH of about 5.6.

Advection Mass motion in the atmosphere; in general, horizontal movement of the air.

Aerosols Solid or liquid particles dispersed in a gas; dust and fog are examples.

Agroclimatology The scientific study of the effects of climate upon crops.

Air mass A large body of air that is characterized by homogeneous physical properties at any given altitude.

Alberta Clipper A fast moving low pressure system originating in Alberta, Canada, and influencing weather in the Midwest.

Anticyclone An area of above-average atmospheric pressure characterized by a generally outward flow of air at the surface.

Aphelion The point on its orbit around the sun where the earth is furthest from the sun.

Atmosphere The mixture of gases that surrounds the earth.

Atmospheric circulation The motion within the atmosphere that results from inequalities in pressure over the earth's surface. When the average circulation of the entire globe is considered, it is referred to as the general circulation of the atmosphere.

Beaufort Wind Scale A system for estimating wind speed named for its inventor, Sir Francis Beaufort.

Blizzard High winds accompanied by blowing snow usually associated with winter cold fronts in middle latitudes.

Blocking	The retardation of eastward moving pressure systems by stagnation of a high pressure system.
Boundary layer	The layer of air above the earth's surface where air motion is influenced by the friction with the earth's surface features. Also called the atmospheric boundary layer or the friction layer.
Carbon monoxide (CO)	A poisonous gas with no color and little odor, formed when there is incomplete combustion of fossil fuels.
Chlorofluorocarbons	A class of compounds whose major ingredients are chlorine, fluorine, and carbon. These compounds are instrumental in the destruction of the stratospheric ozone layer.
Cirrostratus	A thin, whitish veil of cloud that forms at high altitudes.
Cirrus	A feather-like, high-level cloud composed of ice crystals.
Climate	All of the types of weather that occur at a given place over a given time.
Climate model	A numerical representation of a climate system designed to enable an understanding of that system with a view to prediction.
Climatic regime	The annual cycles associated with various climatic elements; for example, the thermal regime is the seasonal patterns of temperature, and the moisture regime is the seasonal patterns of precipitation.
Climatology	The scientific study of climate and climate phenomena.
Cold front	The leading edge of a cold air mass where it displaces a warmer air mass.
Condensation	The change of state from a gas to a liquid.
Conduction	Energy transfer directly from molecule to molecule. It takes place most readily in solids in which molecules are tightly packed.
Continentality	A measure of the remoteness of a land area from the influence of the ocean.
Convection	Mass movement in a fluid or vertical movements in the atmosphere.
Convergent	Moving toward a central point of area; coming together.
Cooling Degree Days (CDD)	The accumulation of temperatures of days above 65°F to provide information about air conditioning costs and amounts of energy needed.
Coriolis force	An apparent force caused by the earth's rotation. It is responsible for deflecting winds clockwise in the Northern Hemisphere and counter clockwise in the Southern Hemisphere.
Cumulonimbus	A vertically developed cloud often topped by an anvil shape. Popularly named anvilhead and often giving rise to severe weather.
Cyclone	Any rotating low pressure system.

Dew point	The temperature at which saturation would be reached if the air mass were cooled at constant pressure without altering the amount of water vapor present.
Diurnal	Having a daily cycle.
Doldrums	An area near the equator of very poorly defined surface winds associated with the equatorial convergence zone.
Doppler radar	A type of radar which detects relative direction of motion and used to locate rotation in clouds and tornadoes.
Drainage basin	A part of the surface of earth that is occupied by a drainage system of rivers and streams.
Drought	A time of unusually low water supply. It may be caused by reduced rainfall or snowmelt, or by low levels of stream flow.
El Niño	An event that is most pronounced in the southern Pacific Ocean that reverses the normal flow of water and wind. Associated with the Southern Oscillation.
ENSO	Acronym for El Niño–Southern Oscillation.
Equinox	About March 21 and September 21 (the Vernal and Autumnal Equinoxes, respectively) the sun crosses the equator and day and night are of equal length.
Evapotranspiration	The total water loss from land by the combined processes of evaporation and transpiration.
Friction layer	The atmospheric layer above the earth's surface in which wind speed and direction are influenced by friction with the earth.
Front	The boundary of intersection between two different air masses.
Fujita scale	A scale which rates the relative severity of tornadoes based on the extent of damage; also known as the F-scale.
Geosynchronous	As applied to a satellite that orbits the earth such that it remains fixed over the same point at the earth's equator.
Global warming	The gradual warming of the earth's climate that is related to both natural and human-caused temperature increase.
Greenhouse Effect	The process by which the heating of the atmosphere is compared to a common greenhouse. Sunlight (short-wave radiation) passes through the atmosphere to reach the earth. The energy reradiated by the earth is at a longer wavelength, and its return to space is inhibited by atmospheric carbon dioxide and water vapor, among other so-called greenhouse gases. This process acts to increase the temperature of the lower atmosphere.
Heat index	A value that describes the combined effect of high temperature and high humidity on the body; also known as the apparent temperature.
Heating Degree Days (HDD)	The accumulation of temperatures of days below 65°F to provide information about heating costs and amounts of energy needed.

Heat island	An area of higher air temperature, compared to the surrounding area, in an urban region.
Heat wave	Any unseasonably warm spell, which can occur any time of the year.
Hurricane	A tropical cyclone that develops and intensifies in the Atlantic Ocean.
Hydrologic cycle	The processes by which water is cycled through the environment. It involves the changes of state of water between solid, liquid, and gas, as well as the transport of water from place to place.
Hyperthermia	The state resulting from a dangerous rise in the body temperature of humans or animals.
Hypothermia	The state resulting from a dangerous fall in the body temperature of humans or animals.
Infrared radiation	Radiation with a wavelength longer than that of visible red light; not visible to the naked eye. Most sensible heat radiated by the earth and other terrestrial objects is in the form of infrared waves.
Inversion	A reversal of the normal decline in temperature with height; for example, as when temperature increases with height.
Isobar	A line on a map or chart connecting points of equal barometric pressure.
Ice Age	A period in earth history in which ice extended far beyond its polar limits.
Indian summer	A time of unusually warm, clear weather in mid or late Autumn after a frost has occurred.
Jet stream	A high-speed flow of air that occurs in narrow bands.
Katabatic wind	Any air blowing downslope as a result of the force of gravity.
Kyoto Protocol	More properly, the Kyoto Protocol to the United Nations Framework Convention on Climate Change is a proposed amendment that was adopted in 1997 in Kyoto, Japan. It commits developed countries to reduce greenhouse gas emissions over a thirty-year period.
La Niña	A time of strong trade winds and low sea surface temperatures in the central and eastern tropical Pacific Ocean. The antithesis of El Niño.
Latent energy	Energy temporarily stored or concealed, such as the heat contained in water vapor.
Mean	A measure of central tendency, the sum of individual values divided by the number of items used to derive the sum.
Median	A measure of central tendency, the value of the middle item when items are arranged according to size.
Meridional circulation	Air flowing in a circulation pattern that is essentially aligned parallel to meridians of longitude.

Mesocyclone	A very large thunderstorm in which rotation has developed. The rotation sometimes results in the formation of a tornado.
Metabolism	The chemical processes that sustain organisms.
Meteorology	Study of phenomena of the atmosphere with a view to understanding and prediction.
Methane (CH$_4$)	A colorless, odorless, and flammable gas. It is among the gases in the atmosphere that absorbs earth radiation quite efficiently.
Microclimate	The climate of small areas, such as a forest floor or small valley.
Monsoon	Atmospheric circulation typified by a change in prevailing wind direction from one season to another.
Nimbostratus	A low dark gray cloud layer.
Occluded front	A warm mass of air trapped when a cold front overtakes a warm front.
Orographic precipitation	Precipitation that results from the lifting of air over some topographic barrier such as a coastline, hills, or mountains.
Ozone (O$_3$)	Oxygen in its triatomic form; highly corrosive and poisonous.
Ozone layer	The layer of ozone, 15½ miles (25 km) above the earth's surface, that absorbs ultraviolet radiation from the sun.
Particulates	Solid particles found in the atmosphere; also known as particulate matter.
Perihelion	The point on its orbit around the sun where the earth is closest to the sun.
Persistence	The continuation of a set of existing conditions.
Photoperiod	The period of each day when direct solar radiation reaches the earth's surface, approximately sunrise to sunset.
Photosynthesis	The process by which sugars are manufactured in plant cells using water and carbon dioxide in the presence of sunlight.
Pluvial	Pertaining to precipitation.
Polar front	The storm frontal zone separating air masses of polar origin from air masses of tropical origin.
Pressure gradient	The amount of pressure change occurring over a given distance.
Radiation	The process by which electromagnetic radiation is propagated through space.
Relative humidity	The ratio of the amount of water present in the air to the amount of water vapor the air can hold, multiplied by 100.
Remote sensing	The acquisition of data or information about an object or phenomena when not in physical contact with it.
Respiration	The physical and chemical processes by which an organism supplies itself with oxygen.
Ridge	An elongated area where pressure is higher than the surrounding region.

Rossby waves	Upper air waves in the middle and upper troposphere of the middle latitudes with wavelengths of 2500–3700 miles (4000–6000 km); named for Carl-Gustav Rossby, the meteorologist who developed the equations for parameters governing the waves.
Saffir-Simpson Scale	A relative scale for categorizing the strength of tropical lows which uses wind velocity, storm surge, and atmospheric pressure.
Sensible temperature	The sensation of temperature the human body feels in contrast to the actual heat content of the air recorded by a thermometer.
Solar constant	The mean rate at which solar radiation reaches the earth.
Solstice	In the Northern Hemisphere, the Summer Solstice occurs about June 21 and the Winter Solstice about December 21 when the sun is highest and lowest in the sky, respectively.
Southern Oscillation	The shifting of atmospheric circulation over the southern Pacific region.
Stationary front	A cold or warm front that has ceased to move; the boundary between two stagnant air masses.
Standard atmosphere	A conventional vertical profile of temperature, pressure, and density within the atmosphere.
Stratosphere	A thermal division of the earth's atmosphere located between the troposphere and the mesosphere. Primary zone of ozone formation.
Stratus	A low cloud that forms a gray layer with a uniform base; a rain cloud, often associated with drizzle.
Sublimation	The transition of water directly from the solid state to the gaseous state without passing through the liquid state.
Subsidence	Descending or setting, as in the air.
Supercell	An extremely large thunderstorm, often reaching as high as the tropopause or sometimes extending into the stratosphere.
Surface ozone	Ozone which forms near the ground as a result of sunlight acting on incompletely combusted hydrocarbons. It occurs most often and most intensely in urban areas of the industrial world.
Synoptic climatology	A study of climatology that relates local and regional climates to atmospheric circulation patterns.
Teleconnections	The link between weather and climate events occurring in widely separated areas of the world.
Terrestrial	Pertaining to the land, as distinguished from the sea or air.
Thermal	A small-scale rising current of warm air.
Thermocline	A vertical temperature gradient that is appreciably greater than temperature gradients in layers above or below; mostly in stratified water.
Thermodynamics	The science of the relationship between heat and mechanical work.

Thunderstorm	A convective cell characterized by vertical cumuliform clouds.
Tornado	An intense vortex in the atmosphere with abnormally low pressure in the center and a converging spiral of high velocity wind.
Transpiration	The process by which water leaves a plant and changes to vapor in the air.
Tropical cyclone	A large rotating low pressure storm that develops over tropical oceans, called a hurricane in the Atlantic Ocean and a typhoon in the Pacific Ocean.
Tropopause	The boundary between the troposphere and stratosphere.
Troposphere	The lower layer of the atmosphere marked by decreasing temperature, pressure, and moisture with height; the layer in which most day-to-day weather changes occur.
Trough	An elongated area of low pressure relative to the surrounding area.
Twilight	The period before sunrise and after sunset in which refracted sunlight reaches the earth.
Ultraviolet radiation	Radiation with a wavelength shorter than that of visible violet light. The invisible ultraviolet radiation is largely responsible for sunburn.
Upslope	Moving uphill, such as a breeze or fog.
Upwelling	A vertical current of water in the ocean. They are common along the eastern sides of the oceans in the middle latitudes.
Virga	A thin veil of rain seen hanging from a thunderstorm but not reaching to the ground. The droplets evaporate before they reach the ground.
Viscosity	The internal friction in fluids that offers resistance to flow; for example, water has a low viscosity, while molasses has a high viscosity.
Vortex	A whirling or rotating fluid with low pressure in the center.
Walker circulation	The atmospheric circulation over the southern Pacific Ocean. It follows the general model of the Hadley cell with westward drift of the trade winds along the surface near the equator and counter-flow aloft.
Warm front	A zone along which a warm air mass is displacing a colder one.
Waterspout	A tornado occurring at sea that touches the surface and picks up water.
Water year	A twelve-month period used in hydrology to correspond to the annual precipitation or flow of streams. In the United States, the water year runs from October 1 to September 30. October 1 is the time of year when water supply in the soil and streams is usually at its lowest.
Wavelength	The linear distance between the crests or the troughs in a successive wave pattern.

Weather	The state of the atmosphere at any one point in time and space.
Windchill	The physiological effects resulting from the combined effect of low temperature and wind.
Wind rose	A class of diagram used to show wind speed and direction.
Zenith	A point in space directly above a person's head.
Zonal circulation	The approximate flow of air along parallel latitudes.

ADDITIONAL REFERENCES AND WORKS CITED

Many of the book references given for chapter 1 (marked *) are general texts that provide excellent coverage of most of the basics of weather and climate.

Chapter 1

*Aguado, E., and J. E. Burt. 2001. *Understanding Weather and Climate.* Upper Saddle River, N.J.: Prentice Hall.

*Gedzelman, S. D. 1980. *The Science and Wonders of the Atmosphere.* New York: Wiley.

*Griffiths, J. F., and D. M. Driscoll. 1986. *Survey of Climatology.* Columbus, Ohio: Merrill.

*Lutgens, F. K., and E. J. Tarbuck. 2001. *The Atmosphere,* 8th ed. Upper Saddle River, N.J.: Prentice Hall.

McKnight, T. L. 1996. *Physical Geography: A Landscape Appreciation,* 5th ed. Upper Saddle River, N.J.: Prentice Hall.

*Miller, A., J. C. Thompson, R. E. Peterson, and D. R. Haragan. 1983. *Elements of Meteorology,* 4th ed. Columbus, Ohio: Merrill.

*Oliver, J. E., and J. J. Hidore. 2002. *Climatology: An Atmospheric Science,* 2nd ed. Upper Saddle River, N.J.: Prentice Hall.

Oliver, J. E. 2005. *Encyclopedia of World Climatology.* New York: Springer.

World Meteorological Organization. 1987. *International Cloud Atlas,* vols. I and II. Geneva: World Meteorological Organization.

Chapter 2

Ahrens, C. D. 1982. *Meteorology Today,* 1st ed. New York: West Publishing.

Bluestein, H. 1999. *Tornado Alley: Monster Storms of the Great Plains.* New York: Oxford University Press.

Fujita, T. T. 1981. Tornadoes and Downbursts in the Context of Generalized Planetary Scales. *Journal of Atmospheric Science* 38: 1511–1534.

Pryor, S. C., and T. Kurzhal. 1997. A Tornado Climatology for Indiana. *Physical Geography* 18: 525–543.

Schmidling, T. W., and J. A. Schmidling. 1996. *Thunder in the Heartland.* Kent, Ohio: Kent State University Press.

Simmons, K. M., and D. Sutter. 2005. WSR-88D Radar, Tornado Warnings, and Tornado Casualties. *Weather and Forecasting* 20: 301–310.

Wilson, J. 2003. *Indiana in Maps: Geographic Perspectives of the Hoosier State.* Indianapolis: Geography Educators' Network of Indiana.

Wilson, J. W., and S. Changnon. 1971. *Illinois Tornadoes.* Urbana, Ill.: Illinois State Water Survey.

Chapter 3

Burroughs, W. J. 1999. *The Climate Revealed.* New York: Cambridge University Press.

Eagleman, J. R. 1990. *Severe and Unusual Weather,* 2nd ed. Lenexa, Kansas: Trimedia.

Geer, I. W. 1996. *Glossary of Weather and Climate.* Boston: American Meteorological Society.

Huschke, R. E. 1959. *Glossary of Meteorology.* Boston: American Meteorological Society.

Morgan, M. D., and J. M. Moran. 1997. *Weather and People.* Upper Saddle River, N.J.: Prentice Hall.

Robinson, P. J. 2001. On the Definition of a Heat Wave. *Journal of Applied Meteorology* 40: 762–775.

Smith, K. 2001. *Environmental Hazards: Assessing Risk and Reducing Disaster,* 3rd ed. New York: Routledge.

Chapter 4

Boubel, R. W., D. L. Fox, D. B. Turner, and A. C. Stern. 1994. *Fundamentals of Air Pollution,* 3rd ed. New York: Academic Press.

Driscoll, D. M. 1985. Human Health, 778–814. In *Handbook of Applied Meteorology,* ed. D. D. Houghton. New York: Wiley.

Turco, Richard P. 2002. *Earth Under Siege: From Air Pollution to Global Change,* 2nd ed. New York: Oxford University Press.

Chapter 5

Eichenlaub, V. L. 1979. *Weather and Climate of the Great Lakes Region.* Notre Dame, Ind.: University of Notre Dame Press.

Henson, Robert. 2002. Cold Rush: Scientists Search for an Index that Fits the Chill. *Weatherwise,* Jan./Feb.

Rothrock, H. J. 1969. An Aid in Forecasting Significant Lake Effect Snows. *ESSA Technical Memorandum* WBTM CR-30. Kansas City: National Weather Service Central Region.

Spar, J. 1957. *The Ways of the Weather.* New York: American Museum of Natural History.

Chapter 6

Dolan, E. F. 1988. *The Old Farmer's Almanac of Weather Lore: The Fact and Fancy behind Weather Predictions, Superstitions, Old-Time Sayings, and Traditions.* Dublin, N.H.: Yankee Books.

Forrester, F. H. 1981. Weather Lore—Facts and Fancies, 278–291. In *1001 Questions Answered about the Weather.* New York: Dover.

Freier, G. D. 1992. *Weather Proverbs: How 600 Proverbs, Sayings, and Poems Accurately Explain Our Weather.* Tucson, Ariz.: Fisher Books.

Lee, A. 1976. *Weather Wisdom.* Garden City, N.Y.: Doubleday.

Horvitz, L. A. 2007. *The Essential Book of Weather Lore.* New York: Penguin.

Sloane, E. 2005. *Eric Sloane's Weather Almanac: A Classic Illustrated Guide to Weather Folklore and Forecasting.* Osceola, Wisc.: Voyageur Press.

Chapter 7

Bhowmik, N. G., et al. 1994. *The 1993 Flood on the Mississippi River in Illinois.* Champaign, Ill.: Illinois State Water Survey.

Bove, M. C. 1998. Impacts of ENSO on United States Tornadic Activity, 199–202. In *Proceedings of the Ninth Symposium on Global Change Studies.* Phoenix, Ariz.: American Meteorological Society.

Changnon, S. A., K. E. Kunkel, and B. C. Reinke. 1996. Impacts and Responses to the 1995 Heat Wave: A Call to Action. *Bulletin of the American Meteorological Society* 77: 1497–1506.

Changnon, S. A., S. D. Hilberg, and K. E. Kunkel. 2000. *El Niño 1997–1998 in the Midwest.* Champaign, Ill.: Illinois State Water Survey.

Changnon, S. A., and D. Changnon. 2005. *The Pre-Christmas 2004 Snowstorm Disaster in the Ohio River Valley.* Champaign, Ill.: Illinois State Water Survey.

Diaz, H. F., and V. Markgraf, eds. 2000. *El Niño and the Southern Oscillation.* New York: Cambridge University Press.

Glantz, M. H. 1996. *Currents of Change: El Niño's Impact on Climate and Society.* New York: Cambridge University Press.

Knowles, J. B., and R. A. Pielke, Sr. 2005. The Southern Oscillation and its Effect on Tornadic Activity in the United States. *Atmospheric Science Paper No. 755.* Fort Collins, Colo.: Department of Atmospheric Science, Colorado State University.

Kunkel, K. E., S. A. Changnon, and J. R. Angel. 1994. Climatic Aspects of the 1993 Upper Mississippi River Basin Flood. *Bulletin of the American Meteorological Society* 75: 811–822.

Lamb, P. J., et al. 1992. *The 1988–1989 Drought in Illinois: Causes, Dimensions, and Impacts.* Champaign, Ill: Illinois State Water Survey.

Chapter 8

Hansen, J., M. Sato, J. Glascoe, and R. Ruedy. 1998. A Common-Sense Climate Index: Is Climate Changing Noticeably? *Proceedings of the National Academy of Sciences* 95: 4113–4120.

Landsberg, H. E. 1981. *The Urban Climate.* New York: Academic Press.

Mackenzie, F. T. 1998. *Our Changing Planet,* 2nd ed. Upper Saddle River, N.J.: Prentice Hall.

Newman, J. E. 1997. Our Changing Climate, 85–99. In *The Natural Heritage of Indiana,* ed. M. T. Jackson. Bloomington: Indiana University Press.

Oke, T. R. 1987. *Boundary Layer Climates.* London: Methuen.

Taylor, D. 2006. Field Work Begins on Project Aimed at Keeping Urban Areas Cooler. *ISU News,* Sept. 6.

Thomas, N. 1995. *Understanding Your Wind Resource.* Washington, D.C.: American Wind Energy Associates.

Visher, S. S. 1945. *The Climate of Indiana.* Bloomington: Indiana University Press.

Chapter 9

Booth, R. K., and S. T. Jackson. 2003. A High-Resolution Record of Late-Holocene Moisture Variability from a Michigan Raised Bog. *The Holocene* 13: 865–878.

Bridgman, H. A., and J. E. Oliver. 2006. *The Global Climate System: Patterns, Processes, and Teleconnections.* New York: Cambridge University Press.

Fleming, A. 1997. Freeze Frame: The Ice Age in Indiana. Indiana Geological Survey Website (http://igs.indiana.edu/Geology/ancient/freezeframe/index.cfm).

Fritts, H. C. 2001. *Tree Rings and Climate.* Caldwell, N.J.: Blackburn Press.

Mock, C. J., J. Mojzisek, M. McWaters, M. Chenoweth, and D. W. Stahle. 2007. The Winter of 1827–1828 over Eastern North America: A Season of Extraordinary Climatic Anomalies, Societal Impacts, and False Spring. *Climate Change* 83: 87–115.

Melhorn, W. N. 1997. Indiana on Ice: Late Tertiary and Ice Age History of Indiana Landscapes, 15–27. In *The Natural Heritage of Indiana,* ed. M. T. Jackson. Bloomington: Indiana University Press.

Stahle, D. W., M. K. Cleaveland, and J. G. Hehr. 1988. North Carolina Climate Changes

Reconstructed from Tree Rings: AD 372 to 1985. *Science* 240: 1517–1519.

Strahler, A. N. 1952. *Physical Geography.* New York: Wiley.

Williams, A. S. 1974. Late-Glacial–Postglacial Vegetational History of the Pretty Lake Region, Northeastern Indiana. *U.S. Geological Survey Professional Paper* 686-B. Washington, D.C.: U.S. Government Printing Office.

Chapter 10

Hidore, J. J. 1996. *Global Environmental Change.* Upper Saddle River, N.J.: Prentice Hall.

Intergovernmental Panel on Climatic Change. 2001. *Climate Change 2001: Synthesis Report. Summary for Policymakers.* Geneva: IPCC.

———. 2007. *Climate Change 2007: Mitigation of Climate Change.* Working Group III contribution to the Fourth Assessment Report of the IPCC. Geneva: IPCC.

———. 2007. *Climate Change 2007: Impacts, Adaptation and Vulnerability.* Working Group II contribution to the Fourth Assessment Report of the IPCC. Geneva: IPCC.

Kling, G. W., K. Hayhoe, L. B. Johnson, et al. 2003. *Confronting Climate Change in the Great Lakes Region: Impacts on Our Communities and Ecosystems.* Cambridge, Mass.: Union of Concerned Scientists; Washington, D.C.: Ecological Society of America.

McGuffie, K., and A. Henderson-Sellers. 2006. *A Climate Modelling Primer,* 3rd ed. New York: Wiley.

Mendelsohn, R., and J. E. Neumann. 2004. *The Impact of Climate Change on the United States Economy.* New York: Cambridge University Press.

Selected Websites

Websites are notorious for having short-term lives and sometimes erroneous information. Those listed below are reliable, and if changes in addresses do occur, the new websites are often provided.

Although the websites listed below parenthetically mention particular chapters, most are multipurpose and contain a wealth of information. For example, the University of Illinois (UICI) website mentions chapter 1, but it contains information applicable to many other chapters.

- http://www.srh.noaa.gov/jetstream/global/jet.htm (accessed July 1, 2008). Part of the weather and climate coverage by the National Weather Service dealing with jet streams (chapter 1).

- http://ww2010.atmos.uiuc.edu/(Gh)/guides/mtr/cld/cldtyp/home.rxml (accessed July 1, 2008). The University of Illinois Weather World 2001 website that deals with clouds. WW2001 provides many other basic weather studies and can be accessed from this website (chapter 1).

- http://www.weather.gov/ (accessed July 1, 2008). The National Oceanic and Atmospheric Administration's National Weather Service website. Weather forecasts, weather safety, and many other useful features are found here (chapter 2).

- http://www.nssl.noaa.gov/edu/safety/tornadoguide.html (accessed July 1, 2008). A well-illustrated guide to many aspects of tornadoes (chapter 2).

- http://www.weather.gov/os/heat/index.shtml (accessed July 1, 2008). The National Oceanic and Atmospheric Administration's National Weather Service website that deals with the Heat Index (chapter 3).

- http://www.drought.gov/ (accessed July 1, 2008). The National Integrated Drought Information System's website for drought information in the United States (chapter 3).

- http://www.usgs.gov/hazards/floods/ (accessed July 1, 2008). The United States Geological Survey's website for flood information. Links to information about other natural hazards are also found here (chapter 3).